Sensors and Signals

Sergey Y. Yurish,
Amin Daneshmand Malayeri
Editors

Sensors and Signals

IFSA International Frequency Sensor Association Publishing

Editors
Sergey Y. Yurish,
Amin Daneshmand Malayeri
Sensors and Signals

ISBN: 978-84-608-2320-9

BN-20150915-XX

BIC: TBM

Contents

Contributors

Hani K. Al-Mohair
School of Electrical & Electronic Engineering, University Sains Malaysia, Pulau Pinang, Malaysia

Marko Beko
Universidade Lusófona de Humanidades e Tecnologias, Lisbon, Portugal, CTS, UNINOVA – Campus FCT/UNL, Caparica, Portugal

Seddik Bri
Materials and Instrumentations Group (MIN), Electrical Engineering Department, High School of Technology, ESTM, Moulay Ismail University, Meknes, 50000, Morocco

Jiri Cerny
Center of Organic Chemistry Ltd., Rybitvi 296, 533 54 Rybitvi, Czech Republic

Beatriz Chicote Gutierrez
University of the Basque Country, Bilbao, Spain

Rui Dinis
DEE/FCT/UNL, Caparica, Portugal, Instituto de Telecomunicações, Lisbon, Portugal

Hadia El-Hennawy
Ain shams University, Electronics and Communication Department, Cairo, Egypt

Mohamed Elkattan
Exploration Division, Nuclear Materials Authority, Cairo, Egypt

Ales Hamacek
University of West Bohemia, Faculty of Electrical Engineering, RICE, Univerzitni 8, 306 14 Plzen, Czech Republic

Aladin Kamel
Ain Shams University, Electronics and Communication Department, Cairo, Egypt

Özlem Karaca Akkan
Izmir Vocational School, Dokuz Eylul University, Izmir 35160, Turkey

Marie Karaskova
Center of Organic Chemistry Ltd., Rybitvi 296, 533 54 Rybitvi, Czech Republic

Lubomir Kubac
Center of Organic Chemistry Ltd., Rybitvi 296, 533 54 Rybitvi, Czech Republic

Jong-Ha Lee
Department of Biomedical Engineering, School of Medicine, Keimyung University, South Korea

Milica Marikj
CTS, UNINOVA – Campus FCT/UNL, Caparica, Portugal

R. M. Megahid
Nuclear Research Centre, Atomic Energy Authority, Cairo, Egypt

Junita Mohamad-Saleh
School of Electrical & Electronic Engineering, University Sains Malaysia, Pulau Pinang, Malaysia

Eduardo Moreno Hernandez
Instituto de Cibernética Matemática y Física. Departamento de Física Aplicada, Vedado, La Habana, Cuba

Stanislav Nespurek
University of West Bohemia, Faculty of Electrical Engineering, RICE, Univerzitni 8, 306 14 Plzen, Czech Republic

Iñigo Javier Oleagordia Aguirre
Department of Electronic Technology. E.U.I.T.I. Bilbao, University of the Basque Country, Bilbao, Spain

David Rais
Institute of Macromolecular Chemistry, AS CR, 162 06 Praha, Czech Republic

Jan Rakusan
University of West Bohemia, Faculty of Electrical Engineering, RICE, Univerzitni 8, 306 14 Plzen, Czech Republic

Jan Reboun
University of West Bohemia, Faculty of Electrical Engineering, RICE, Univerzitni 8, 306 14 Plzen, Czech Republic

Ahmed Rhbanou
Department of Mathematics, FSM, Moulay Ismail University, Meknes, 50000, Morocco

Mohamed Sabbane
Department of Mathematics, FSM, Moulay Ismail University, Meknes, 50000, Morocco

Radosveta Sokullu
EEE, Faculty of Engineering, Ege University, Bornova Campus, 35100, Izmir, Turkey

Fouad Soliman
Exploration Division, Nuclear Materials Authority, Cairo, Egypt

Jiri Stulik
University of West Bohemia, Faculty of Electrical Engineering, RICE, Univerzitni 8, 306 14 Plzen, Czech Republic

Shahrel Azmin Suandi
School of Electrical & Electronic Engineering, University Sains Malaysia, Pulau Pinang, Malaysia

Slavisa Tomic
ISR/IST, Lisbon, Portugal

Milan Tuba
Megatrend University, Faculty of Computer Science, Belgrade, Serbia

Chapter 1

Virtual Instrumentation with Embedded Control Applied to the Analysis of Ultrasonic Signals in Metallic Structures

Beatriz Chicote Gutierrez, Eduardo Moreno Hernandez, Iñigo Javier Oleagordia Aguirre

1.1. Introduction

Due to the demonstration that the Lamb waves [1] are an effective method for the detection and location of defects or damages in materials, these have led to a growing interest in applications that use this type of waves in non-destructive testing [3] because they turn out to be a topic of great interest both in engineering and science.

Lamb waves [1] are dispersive, which means that their phase and group velocities vary with the frequency and therefore they have an infinite number of propagation modes. However, selecting the frequency and the windowed excitation of a pulse train it is possible to achieve that when these waves propagate trough a material their particles move just in two different ways: symmetric Lamb waves and antisymmetric Lamb waves.

For this reason the objective of this experiment is to verify the mode of propagation of these waves which are generated by the ultrasonic equipment on a steel plate of 5 mm, through the development of a virtual instrument with which is performed the acquisition, control, processing and monitoring of the signals obtained.

This chapter is organized as follows. Section 1.2 gives the description of the hardware used; Section 1.3 describes briefly the programming environment used for the comparison of the theoretical and practical dispersion curves through Fourier algorithms modules and its previous acquisition and signal processing; In Section 1.4 results of the experiment; Conclusions are given in Section 1.5.

1.2. Description of the Hardware

To perform this experiment an ultrasonic device is used, which is formed by:

- Pairs of piezoelectric transducers [6] acting one of them as a transmitter and the other one as a receiver (Panameterics - NDT V194; 1 MHz / 1,125" - 780681 and Panameterics - NDT V191; 0.5 MHz / 1,125" - 720685) which are placed on a steel plate of 1100 mm × 1100 mm, with a thickness of 5 mm which serves as mean of propagation for the waves as shown in Fig. 1.1.

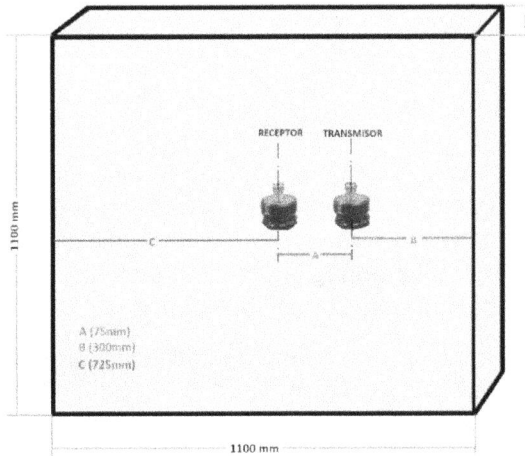

Fig. 1.1. Transducers positioning in the steel plate.

- An analog module, whose function is to excite the piezoelectric transducers capable of emitting ultrasonic pulses.

- A digital module to generate and update the necessary control signals for the operation of the analog module, using reconfigurable logic.

In addition, all these functions can be updated via USB, which is contained in the digital module and connected to a PC, whose operating system is Windows 7, which in turn contains a virtual instrument developed in LabVIEW 2012 (*National Instruments*) programming environment.

14

The Fig. 1.2 shows a block diagram concerning the ultrasonic equipment.

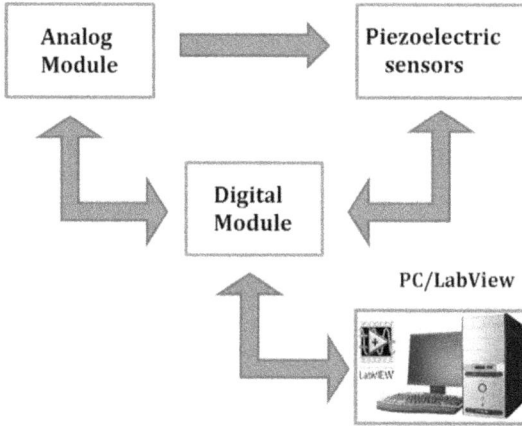

Fig. 1.2. Ultrasonic equipment block diagram.

1.3. Description of the Software

This section describes briefly the programming made for the generation of a virtual instrument with which the user can generate and update the signals of control (frequency, number of pulses, gain, PRF (pulse repetition frequency)) needed for the creation of ultrasonic signals and from the selection of a frequency, using the Fourier algorithms, graphing the frequency-velocity relationship (dispersion curves).

LabVIEW (Laboratory Virtual Instrumentation Engineering Workbench) is a platform and development environment to design systems, with a graphical and visual programming language. LabVIEW is a graphical language developed for scientists and engineers and quite different in the way the code is constructed and saved.

To generate and update the signals above mentioned the digital module communicates with the PC and vice versa using the nodes "Call library function node", whose function is to link functions of shared libraries (interconnection with codes created in other languages, which in this case is VHDL code and the code of the virtual instrument, which is LabVIEW), achieving at first the configuration of the communication port as it can be seen in Fig. 1.3 and secondly the update (reading or writing) of the register, as it can be seen in Fig. 1.4).

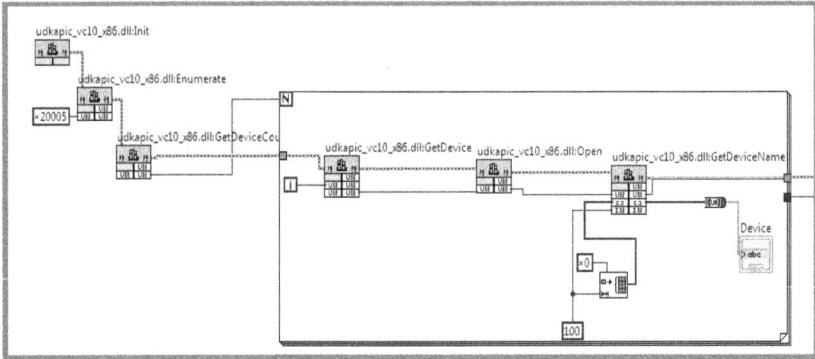

Fig. 1.3. Configuration of the communication port.

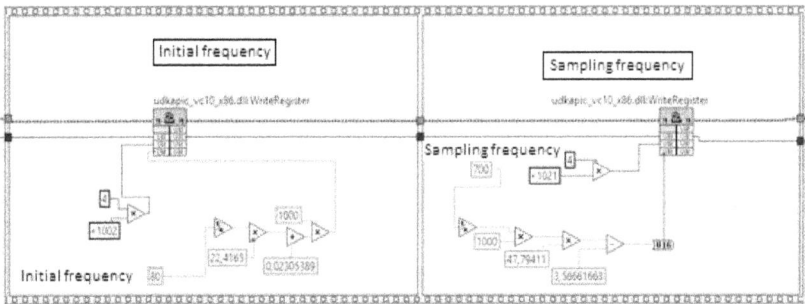

Fig. 1.4. Update (writing) of the system registers.

Then these data are processed so that they can be displayed on the front panel of the virtual instrument as shown in Fig. 1.5.

To obtain this signals is necessary to adjust its control signals which are: frequency, gain, number of pulses and repetition frequency. It can be handled through the LabVIEW Virtual Instrument from the buttons in the front panel that LabVIEW offers.

This way, it can be obtained a panoramic signal (Fig. 1.5.1) of the ultrasonic signal and changing the position of the cursors of this signal it is possible to make a zoom of the signal (Fig. 1.5.2).

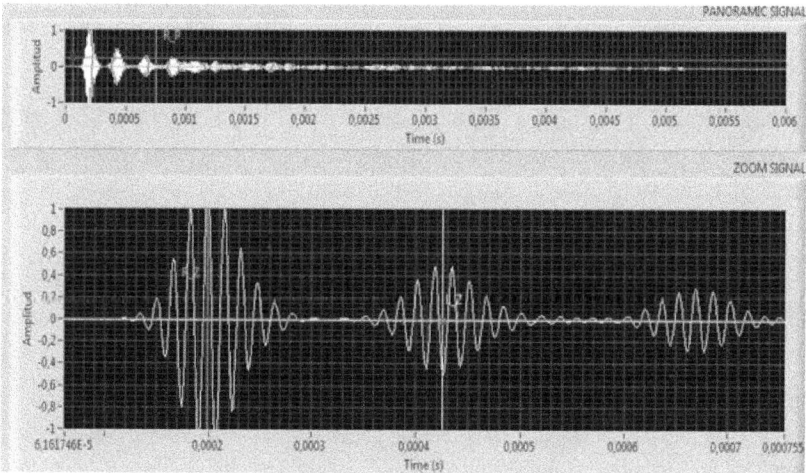

Fig. 1.5. Virtual instrument front panel.

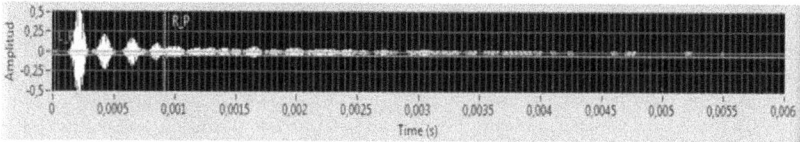

Fig. 1.5.1. Panoramic Signal.

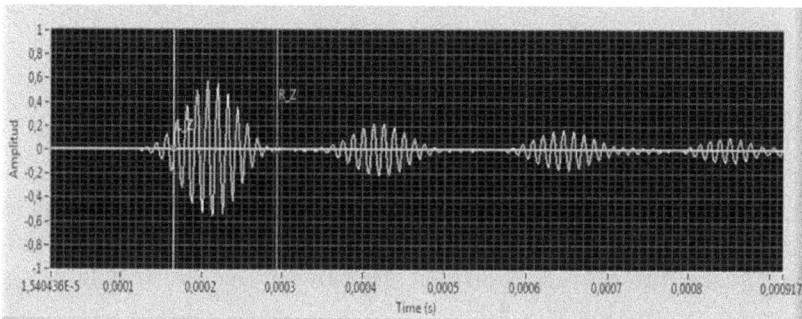

Fig. 1.5.2. Zoom signal.

1.3.1 Fourier

On the other hand, for the graphical representation of frequency an velocity the following Fourier algorithms are necessary:

1.3.1.1. FFT (Fast Fourier Transformation) [4]

In the Fig. 1.6 it is shown the programming of the sub VI FFT_SUB_1 of the FFT by which a set of N samples captured {x[n]} are shown in the frequency domain X[k] using the expression:

$$X[k] = \sum_{n=0}^{N-1} x[n] \cdot e^{-j \cdot \frac{2 \cdot \pi}{N} \cdot k \cdot n} \qquad (1.1)$$

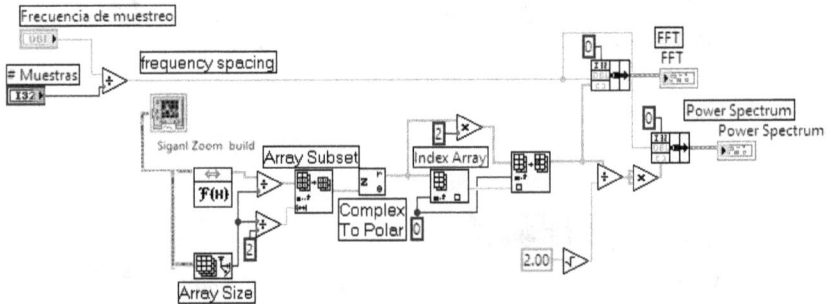

Fig. 1.6. Obtaining of the spectrums of the signal (FFT).

As a theoretical application of the FFT Fig. 1.7 shows the temporal representation of 1000 samples of a square signal of 50 Hz, with an amplitude between -5 V and 5 V and modeled in time domain.

1.3.1.2. STFT (Short-Time Fourier Transform)

With the STFT VI that LabVIEW offers a time-frequency analysis is obtained and is possible to study the evolution of the spectrum in time. From the STFT and the distance between the two transductors is possible to calculate the Slowness, so that the slowness is the inverse of the velocity. (Shown in Fig. 1.8.)

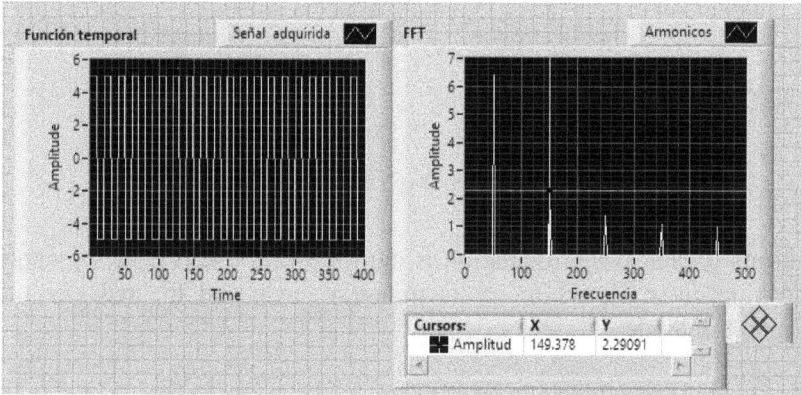

Fig. 1.7. Representation in time and frequency domain.

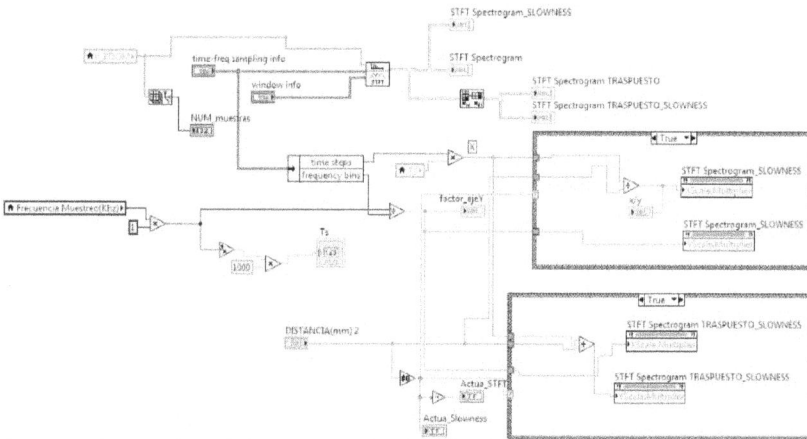

Fig. 1.8. Obtaining the STFT and slowness.

1.3.2. Software Dispersion Curves

To obtain the theoretical dispersion curves a free software is used in which the features of the plate are introduced. In this case are material: steel and thickness: 5 mm. The following dispersion curves are obtained:

- Group velocity: Represents the velocity at which the variations in the shape of the wave amplitude (also named modulation) propagate in the space and is represented by:

19

$$v_g = \frac{\partial_W}{\partial_K},$$ (1.2)

where:

Vg = Group velocity;
W= Wave angular velocity;
K=Wave number.

And its graphical representation is the following one (Fig. 1.9), where the blue curve shows the antisymmetric mode and the red one the symmetric mode.

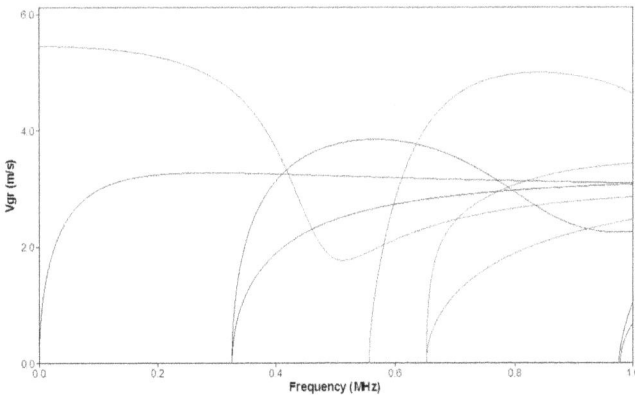

Fig. 1.9. Group velocity.

- Wave phase velocity: Is the velocity at which the phase of any component in frequency of a wave is propagated and is represented by:

$$v_p = \frac{w}{k}$$ (1.3)

Being its graphical representation the one shown in Fig. 1.10.

- Group slowness: The inverse to the group velocity (shown in Fig. 1.11).

- Phase slowness: The inverse to the phase velocity (shown in Fig. 1.12).

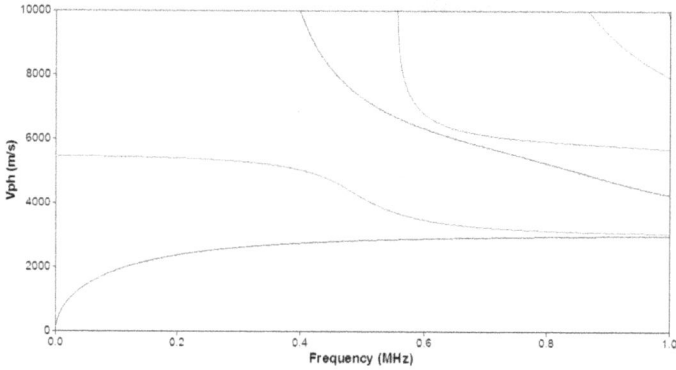

Fig. 1.10. Wave phase velocity.

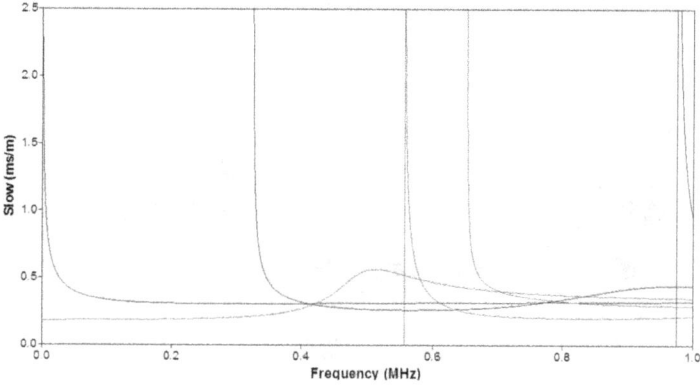

Fig. 1.11. Group slowness.

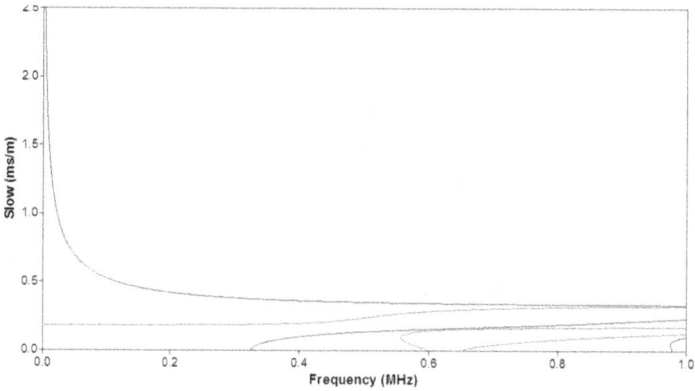

Fig. 1.12. Phase slowness.

21

1.4. Experimental Results

The virtual instrument that has been developed using LabVIEW provides of all the functionalities that are needed to carry out this experiment. The system has been proved in a laboratory with a steel plate of 5 mm and the following results have been obtained:

For an initial frequency of 80 kHz and a gain of 5 V the following signal has been obtained (Fig. 1.13) to which it has been applied a band-pass filter.

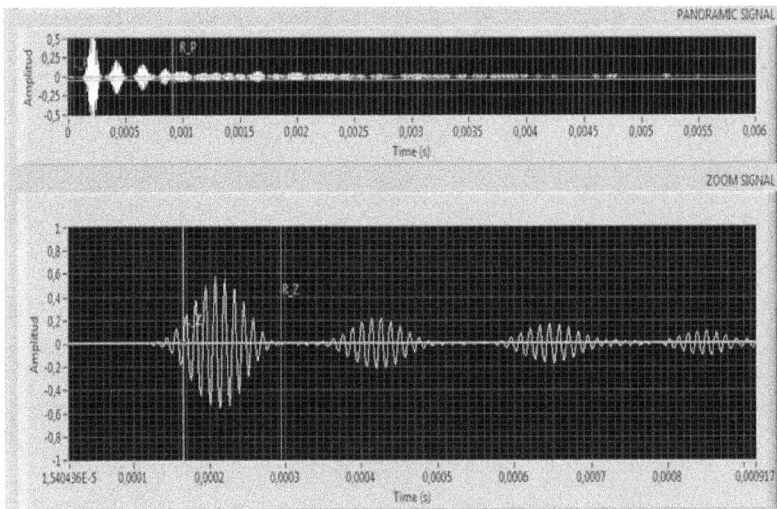

Fig. 1.13. Front Panel. Signal of 80 kHz with band-pass filter.

In the previous signal the first three initial pulses (being the first signal the direct mode) can be appreciated being all others echoes of the signal concerning to the way of propagation of the waves between transmitter and receiver (shown in Fig. 1.14).

In this signal is calculated the STFT and the corresponding Slowness. For that, is necessary to remove the band-pass filter in order to observe the signal in all the range of frequencies provided by the transducers (shown in Fig. 1.15).

For the representation of the Slowness is necessary to give a value of distance, which corresponds with the distance of the direct propagation

22

wave (blue arrow in Fig. 1.14), it can be seen in the steel plate in the Fig. 1.16.

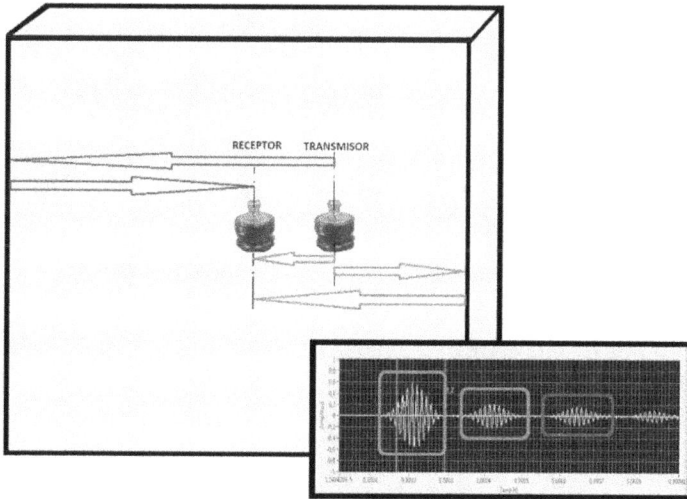

Fig. 1.14. Ways of propagation of the waves in the steel plate.

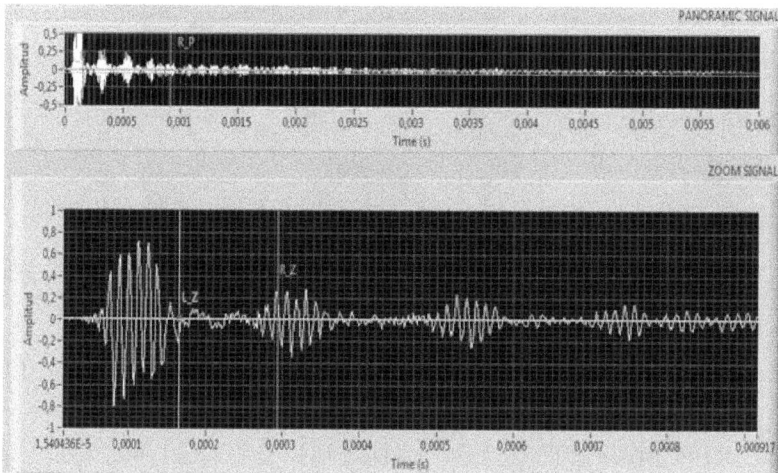

Fig. 1.15. Front Panel. Signal of 80 kHz without band-pass filter.

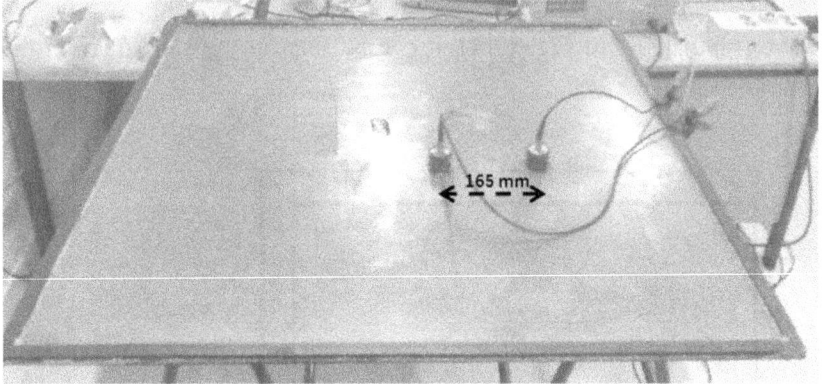

Fig. 1.16. Steel plate. Direct propagation distance.

And with this distance we can obtain the STFT and Slowness, as we can see in Fig. 1.17.

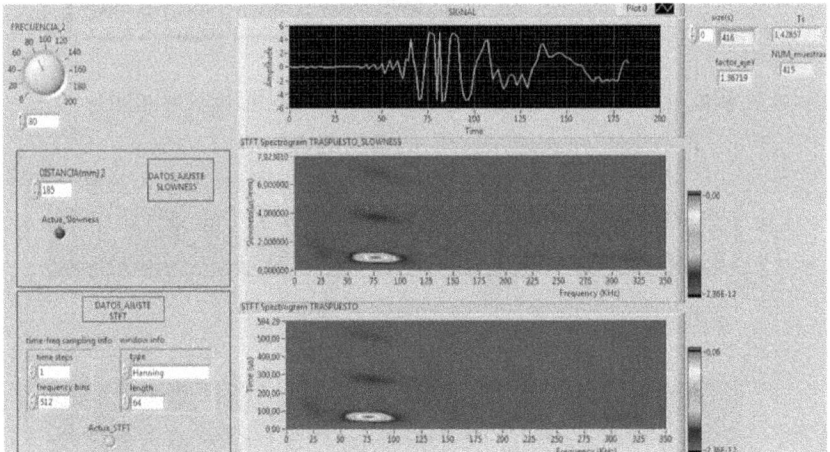

Fig. 1.17. Graphic representation of STFT and Slowness. 80 kHz.

In the same way, for frequencies of 60 kHz (Fig. 1.18), 40 kHz (Fig. 1.19), 30 kHz (Fig. 1.20) and 20 kHz (Fig. 1.21) the following results are obtained:

24

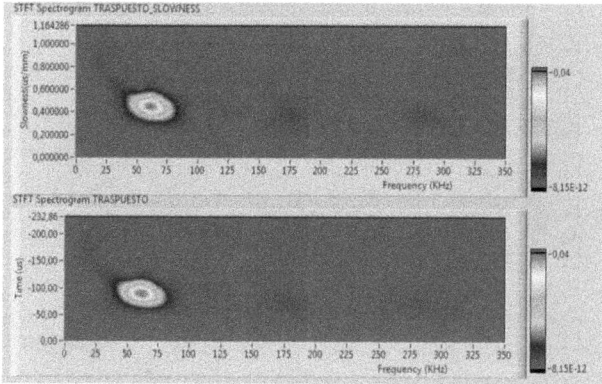

Fig. 1.18. Graphic representation of STFT and Slowness. 60 kHz.

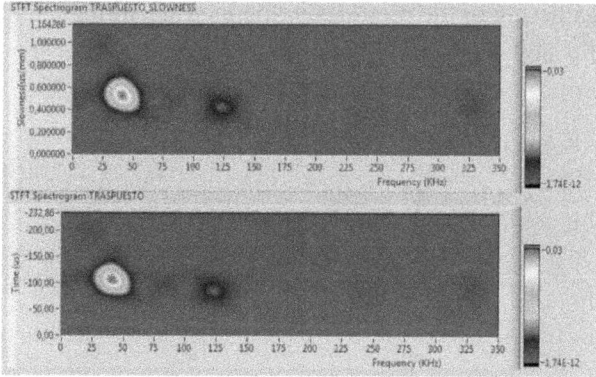

Fig. 1.19. Graphic representation of STFT and Slowness. 40 kHz.

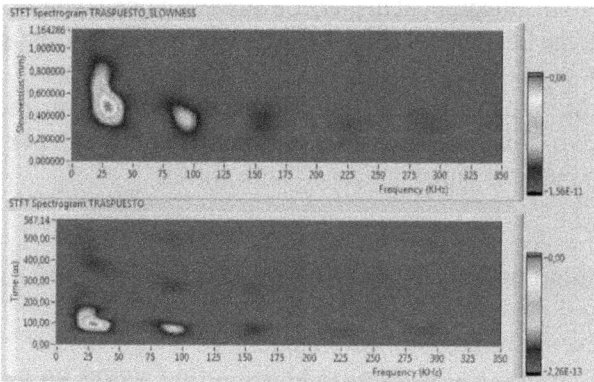

Fig. 1.20. Graphic representation of STFT and Slowness. 30 kHz.

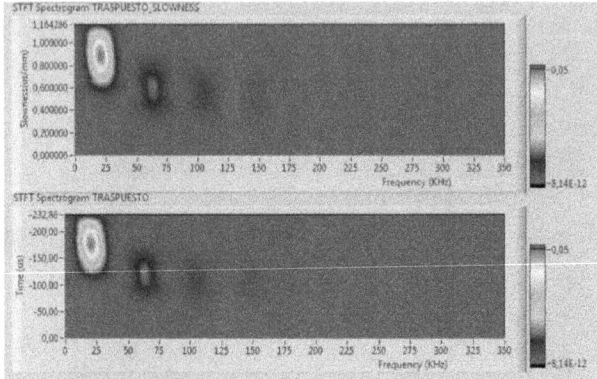

Fig. 1.21. Graphic representation of STFT and Slowness. 20 kHz.

Making a gathering of the previously obtained signal for each frequency we obtain the signal shown in Fig. 1.22.

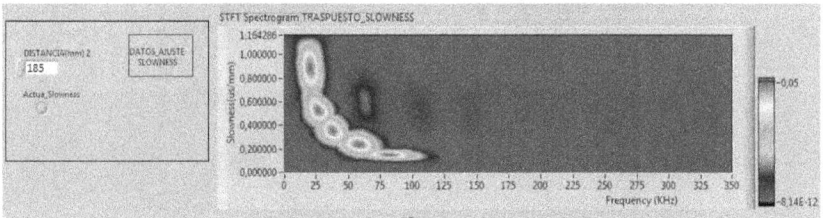

Fig. 1.22. Slowness calculation for different frequencies.

According to the theoretical dispersion curve obtained with the Disperse Software in the Section 1.3 (Fig. 1.12) and the practical dispersion curve obtained in the Section 1.4 (Fig. 1.22) it can be seen that the practical results are like the theoretical (Fig. 1.23) and in conclusion, the particles of this waves that are transmitted through the steel plate of this experiment move as antisymmetric Lamb waves.

1.5. Conclusion

This chapter describes the development of a system that enables to contrast the theoretical and practical dispersion curves on a steel plate with the characteristics previously explained.

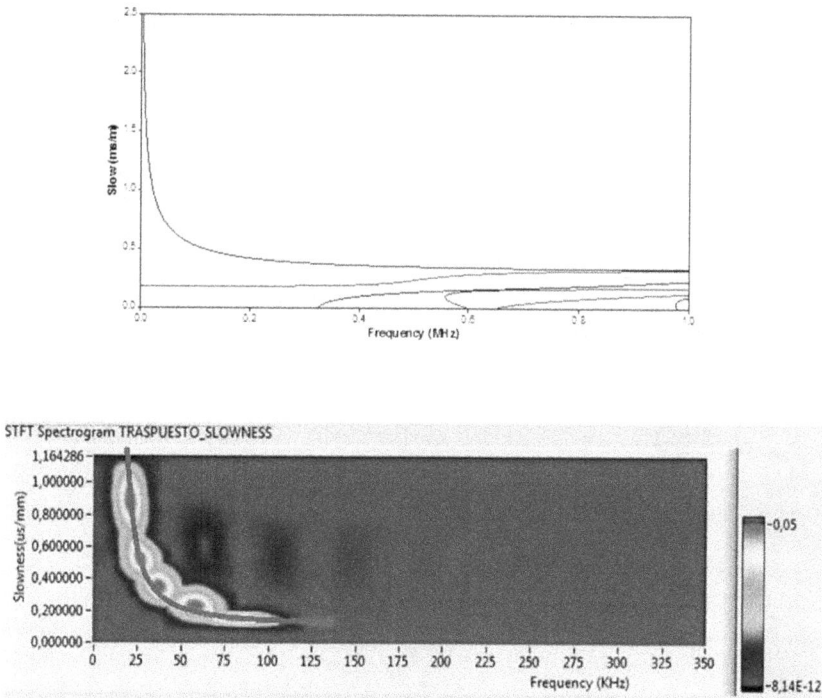

Fig. 1.23. Theoritical-practical comparation.

It has been developed a full system that enables non-intrusive testing based on ultrasonic signals applied in a steel plate. This system enables the integration of hardware and software through the development of a virtual instrument (VI). Also, it has a bigger flexibility compared with other systems. With this system it has been achieved an analysis in time and frequency domain and software modules had been integrated to the generation of the signals. (i.e. Fig. 1.15).

References

[1] Bao. J., Lamb wave generation and detection with piezoelectric wafer active sensors, PhD Thesis, University of South Carolina, 2003.
[2] A. G. Cuc, V. Giurgiutiu, Propagation of guided Lamb waves in bonded specimens, *Proceedings of SPIE*, 2006.
[3] L. &. F. L. Demer, Lamb Wave Techniques in Nondestructive Testing, *International Journal of Nondestructive Testing*, 1969, pp. 251-283.

[4] H. Sorense, Efficient computation of the short-time fast Fourier transform, in *Proceedings of the International Conference on Acoustics, Speech, and Signal Processing, (ICASSP-88),* 1988, New York, NY, 1988.

[5] M. Brook, General Characteristics of Nondestructive Testing (NTD) Methods, *Wiley-IEEE Press*, 2012, p. 12.

[6] M. Indou, Non-destructive inspection of concrete structures by ultrasonic sensor taking account of sensor characteristics, in *Proceedings of the SICE Annual Conference*, Fukui, Japan, 2003.

Chapter 2. Humidity Sensors Based on Substituted Phthalocyanines and either TiO₂
or Carbon Black Nanolayers Modified with them and on the Sandwich Double Layer
Structures, Consisting of Hydrophilic and Hydrophobic Phthalocyanine Derivatives

Chapter 2

Humidity Sensors Based on Substituted Phthalocyanines and either TiO_2 or Carbon Black Nanolayers Modified with them and on the Sandwich Double Layer Structures, Consisting of Hydrophilic and Hydrophobic Phthalocyanine Derivatives

Jan Rakusan, Ales Hamacek, Jan Reboun, Stanislav Nespurek,
Jiri Stulik, Marie Karaskova, Jiri Cerny, David Rais and Lubomir Kubac

2.1. Introduction

Over many years humidity sensors have attracted considerable attention of engineers due to their useful applications in industry and environmental monitoring. [1, 2]. Sensors based on Al_2O_3 are very often used [3]. Cheap humidity sensors are often based on thin films of polymers consisting of saturated main chains and ionizable side groups [4, 5]. Sensitivity to humidity is based on the dissociation of side groups of the type A^+B^-, e.g., COO^-Na^+ and $SO3^-Na^+$. The films can be prepared by cold technologies, like spin coating polymer solutions, casting, doctor blading, etc. Sensors based on ceramics have often an insufficient sensitivity over wide humidity range, bad reversibility, drifts in base resistance due to the chemisorption of water molecules, etc. Our interest is the developing of fast and sensitive humidity sensors based on derivatives of phthalocyanine. We described the hybrid humidity sensor based on TiO_2 nanolayer modified with sodium salt of sulfonated hydroxyaluminium phthalocyanine (AlOHPcSNa) [6]. We also described new types of hybrid humidity sensors based on Carbon nanoparticles and sodium salt of sulfonated Nickel phthalocyanine (NiPcSNa) [7]. The idea was based on the fabrication of sensors with nanolayers of Carbon nanoparticles modified with NiPcSNa, which is deposited either beneath the carbon layer or on the surface of it. We also developed a new humidity sensor sensitive over wide humidity range (20 – 90 % RH), based on active layer of sodium salt of sulfonated Nickel phthalocyanine (NiPcSNa), covered with the protective layer of sulfonamidic nickel phthalocyanine (NiPc(SA)) [8, 9].

2.2. Humidity Hybrid Sensor Based on TiO$_2$ Nanolayer Modified with Sulfonated Hydroxyaluminium Phthalocyanine

2.2.1. Materials

TiO$_2$ nanoparticles were applied in the form of paste (Solaronix, Switzerland, Type: HT/SP). Sodium salt of sulfonated AlOH phthalocyanine, AlOHPc(SO3Na)$_{2-3}$ ((AlOHPcSNa) in further text) was synthesized in the way described below. Its chemical structure is shown in Fig. 2.1. The material used in this work consisted mainly of di- (x = 2) and tri- (x = 3) sulphonated molecules.

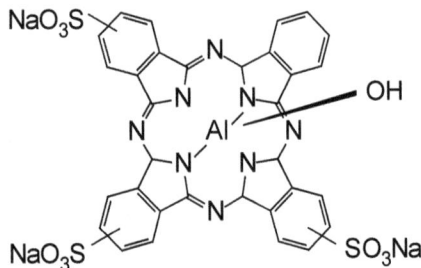

Fig. 2.1. Sodium salt of sulfonated AlOH phthalocyanine AlOHPcSNa.

Sulfonated hydroxyaluminium phthalocyanine AlOHPcS was synthesized from hydroxyaluminium phthalocyanine (AlOHPc). The sulfonation process was carried out in the four necked glass flask, which was heated in an oil bath, equipped with reflux cooler, agitator, thermometer and dosing funnel. 300 g of fuming sulfuric acid was charged into the flask and subsequently 30 g of AlOHPc was gradually charged through the dosing funnel into the agitated fuming sulphuric acid. The mixture was agitated at 30 °C until all the AlOHPc was fully dissolved. After that the dosing funnel was removed and an inlet of nitrogen was installed instead of it. The process of sulfonation was then

Chapter 2. Humidity Sensors Based on Substituted Phthalocyanines and either TiO₂
or Carbon Black Nanolayers Modified with them and on the Sandwich Double Layer
Structures, Consisting of Hydrophilic and Hydrophobic Phthalocyanine Derivatives

carried out under the nitrogen blanket. The reaction mixture was agitated and heated up to temperature from 115 °C to 125 °C and subsequently kept for 30 minutes according to the target level of the sulfonation. After the finishing of the sulfonation process, the reaction mixture was slowly cooled down to the laboratory temperature and subsequently charged with a dropping funnel into the vigorously agitated mixture of 2000 g of ice and 1000 g of water. The water suspension was then filtered with a Buchner funnel, the filter cake was washed with distilled water until no sulfate anions were detectable in the filtrate. Thoroughly washed filter cake of AlOHPcS was dried in laboratory dryer at the temperature of 105 °C until the constant weight. The contents of particular sulfonated (AlOH)Pcs were determined by HPLC. The lower sulfonated mostly mono- and di-sulfonated ones, were obtained when 5 % fuming sulfuric acid was used at the temperature of 115 °C; the contents of mono-sulfonated and di-sulfonated AlOHPcs, as determined by high pressure liquid chromatography (HPLC), were 35.7 and 53.0 %, respectively. The highly sulfonated AlOHPcs, mostly di- and tri-sulfonated, were obtained when 6 % fuming sulfuric acid was used at the temperature of 125 °C. The contents of di-sulfonated and tri-sulfonated AlOHPcs, as determined by HPLC, were 28.3 % and 54.5 %, respectively.

AlOHPcSNa was prepared by dissolving of dry AlOHPcS in water at pH 11 which level was kept by gradual addition of 10 % NaOH water solution. The solution obtained in this way was dried in rotary laboratory vacuum evaporator. The dry product, AlOHPcSNa, contained 28.3 % di, 54,5 % tri sulfonated species, according to HPLC analysis.

2.2.2. TiO₂ Thin Films Preparation and Modification of them with AlPcSNa

Thin films of TiO₂ were prepared from the Solaronix paste by screen printing on glazed ceramic substrates with gold interdigital electrodes (electrode gap was 50 μm). After the deposition the layers were annealed at 450 °C, 60 min. The higher annealing temperatures (up to 500 °C) and longer annealing time (up to 12 h) did not influence the sensor properties too much. The microscopic image of TiO₂ layer (laser confocal mode magnification 7200×) is shown in Fig. 2.2. Subsequently the TiO₂ layer was modified with sodium salt of AlOHPcSNa, containing 28.3 % of di-sulfonated and 54.5 % of tri-sulfonated product. The AlPcSNa was deposited on the TiO₂ layer from an alkali water solution by dipping method for 60 min and after dipping drying at 80 °C. Concentration of AlPcSNa was 200 mg per 1 l of water.)

Fig. 2.2. Microscopic image of TiO$_2$ layer – Solaronix HT/SP paste, magnification 7200 times, laser confocal mode.

2.2.3. Response Detection

The response on humidity exposure was monitored by means of impedance change and detected as the change of frequency (f) obtained in impedance-frequency converter.

2.2.4. Results and Discussion

The dependences of the frequency (f) on relative humidity (RH) are shown in Fig. 2.3. From the figure it follows that TiO$_2$ layer seems to be good humidity sensor, linear in the semilogarithmic plot of log f vs. RH (curve 2). However, the response of the sensor is not constant in long time due to the sorption of various pollutants from environment as it follows from Fig. 2.4, cycles set 2, full line. We were not able to compensate this drift.

Chapter 2. Humidity Sensors Based on Substituted Phthalocyanines and either TiO₂
or Carbon Black Nanolayers Modified with them and on the Sandwich Double Layer
Structures, Consisting of Hydrophilic and Hydrophobic Phthalocyanine Derivatives

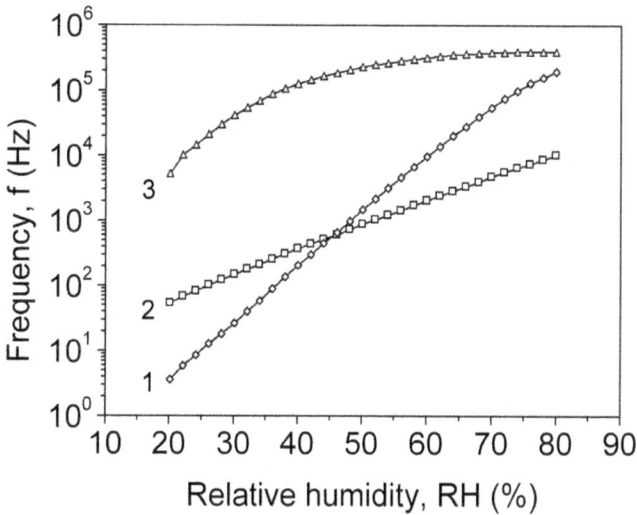

Fig. 2.3. The dependences of the detection frequency on relative humidity, curve 1 – layer TiO₂ annealed at 450 °C for 60 min with adsorbed AlOHPcSNa; curve 2 – TiO₂ layer annealed at 450 °C for 60 min; without AlOHPcSNa; curve 3 – film of AlOHPcSNa.

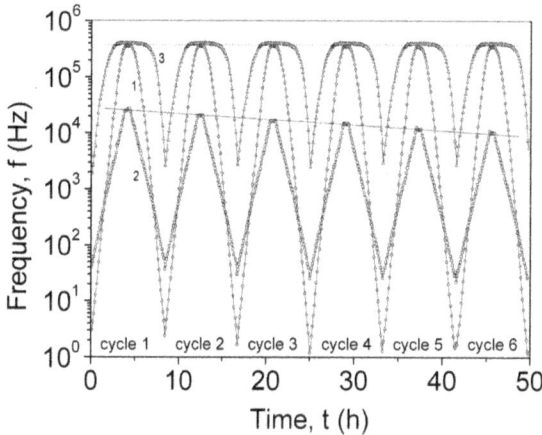

Fig. 2.4. The dependences of the detection frequencies on time at 40 °C. Testing cycle consists of a stepwise humidity change to and from 20 and 90 % RH in about 2 % RH steps.

33

Thus the TiO_2 layer was modified by phthalocyanine AlPcSNa. It must be noted that Sodium salt of AlPcSNa alone is not suitable for the sensor fabrication, as follows from Fig. 2.3, curve 3:

(i) The course of the log f vs. RH dependence was not linear and frequency shows a saturation for humidity higher than about 50 % RH.

(ii) The film is not stable at higher humidity and is washed out from the electrode system due to its water solubility.

The advantage of the hybrid sensor is that the responses are stable in time as it follows from Fig. 2.4, cycles set 1, dashed line. The dependence of the log f vs. RH of the TiO_2 with adsorbed AlPcS is shown in Fig. 2.2, curve 1: The response is linear in log f vs. RH coordinates. The weak sublinearity above 75 % RH does not influence the accuracy of the humidity detection too much. The device sensitivity $S = \Delta \ln f / \Delta RH$ (f in Hz and RH in %) was increased from $S = 0.087$ to $S = 0.19$.

The courses in the plot illustrate the behavior of the three types of sensor during six cycles.

Cycles set 1 – layer TiO_2 annealed at 450 °C for 60 min. with adsorbed phthalocyanine AlOHPcSNa.

Cycles set 2 – TiO_2 layer annealed at 450 °C for 60 min. without AlOHPcSNa

Cycles set 3 – film of AlPcSNa is not stable at higher humidity and is washed out from the electrode system due to its water solubility.

AlPcSNa is very stable organic compound. The substitution with SO_3Na groups makes it hydrophilic. The ability of the substituent to dissociate results in the improved electrical (ionic) conductivity. It allows to increase the sensitivity of TiO_2 layer to humidity.

34

Chapter 2. Humidity Sensors Based on Substituted Phthalocyanines and either TiO₂
or Carbon Black Nanolayers Modified with them and on the Sandwich Double Layer
Structures, Consisting of Hydrophilic and Hydrophobic Phthalocyanine Derivatives

2.3. Phtalocyanine Hybrid Humidity Sensors Based on Carbon Nanoparticles and Sodium Salt of Sulfonated Nickel Phthalocyanine

2.3.1. Materials

2.3.1.1. Sodium Salt of Sulfonated Nickel Phthalocyanine

Sodium salt of sulfonated Nickel phthalocyanine (NiPcSNa) was synthesized in the way as described lower.

2.3.1.2. Thin Carbon Films Preparation

The carbon film was applied on the on glazed ceramic substrates with gold interdigital electrodes (electrode gap was 25 µm) by the method of physical vapour deposition (PVD). To ensure the reproducibility of mechanical and electrical parameters of the carbon layer, the following procedure was applied: The basic substrate was cleaned in ultrasonic bath and dried. After that the substrate was placed into a vacuum chamber of Carbon sputtering machine. The chamber was then evacuated and filled by argon for three times. The pressure after the last cycle was around 4×10^{-2} mbar. The carbon layer was deposited on the substrate by PVD method at 6 V. DC current flowing through the carbon rod, which was situated above the substrates, was 150 A. The thickness of the carbon layer was controlled by deposition time. After 1 s deposition the layer was 6 nm thick, after 2 s deposition 11 nm thick. The substrate covered with one carbon layers without NiPcSNa was used as control unit of humidity sensitivity.

2.3.1.3. Double Layer NiPcSNa/Carbon Humidity Sensors

The humidity sensor was prepared on ceramic substrate, containing 99 % of alumina, with interdigital Au electrodes having isolation gaps 25 µm wide.

The electrodes were made by lift off method. This method is based on cold sputtering of metal in vacuum on the substrate with photoresist pattern. Electrodes were made by the sequential deposition of Au layers. The thickness of the electrodes was about 400 nm.

Two methods were used for the preparation of double layer NiPcSNa/Carbon humidity sensors.

(i) Sodium salt of sulfonated Nickel phthalocyanine NiPcSNa was deposited on the substrate with interdigital electrodes from water solution concentration of 3 % by spin-coating. The rotation speed of spin-coater was 3000 rpm. Subsequently the NiPcSNa layer was covered by carbon nanolayer, 6 nm thick by the PVD method described above sensor (NP-C- 2T, see Table 1.1).

(ii) NiPcSNa layer was deposited by spin-coating from water solution, concentration of 3 %, on carbon nanolayer (6 or 11 nm thick) covering the interdigital electrode system on the substrate (sensors C-NP-1T and C-NP-1S, see Table 1.1).

The substrate covered with one carbon layer 6 nm without NiPcSNa was used as control unit of humidity sensitivity (sensor CT1, see Table 1.1).

Table 1.1. Prepared and measured humidity sensors used for testing in climatic chamber.

Sensor number	Components	Structure	Remark
CT1	C only	C layer only	C layer 6 nm thick
C-NP-1T	NiPcSNa +C	NiPcSNa on C layer	C layer 6 nm thick
C-NP-1S	NiPcSNa +C	NiPcSNa on C layer	C layer 11 nm thick
NP-C-2T	NiPcSNa +C	C layer on NiPcSNa	C layer 6 nm thick

2.3.2. Measurement of Sensor Characteristics

The measurements of properties of double layer NiPcS/C humidity sensors and reference sensor mentioned in Table 1.1 were carried out in a climatic chamber with controlled temperature and humidity. The measurements were fully automatic in humidity and temperature ranges from 20 to 90 % RH and from 20 to 50 °C, respectively. Humidity was increased stepwise, 0.5 % RH per min. The measurement cycle is shown in Fig. 2.5.

Electrical parameters of samples were measured by RLC bridge Agilent E4980A. The results of the measurements are shown in Figs. 2.6 – 2.9.

Chapter 2. Humidity Sensors Based on Substituted Phthalocyanines and either TiO₂
or Carbon Black Nanolayers Modified with them and on the Sandwich Double Layer
Structures, Consisting of Hydrophilic and Hydrophobic Phthalocyanine Derivatives

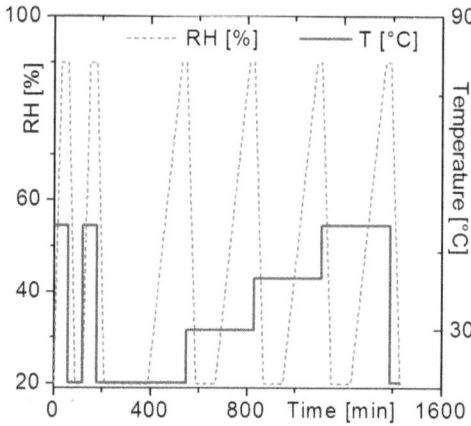

Fig. 2.5. Humidity and temperature cycles during the measurements.

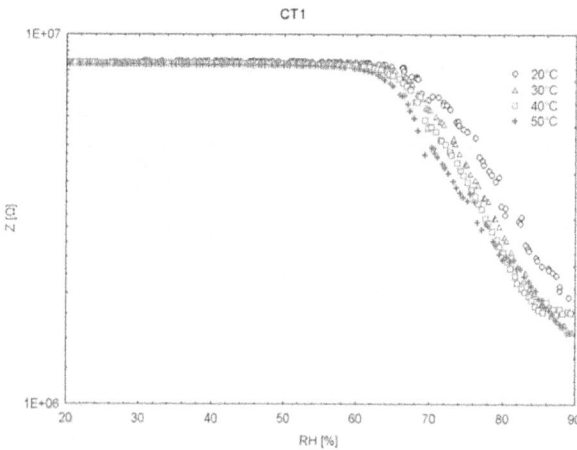

Fig. 2.6. Humidity dependence of impedance of reference humidity sensor
CT1 (only C nano layer 6 nm).

Stability test was performed using the cycles 30 °C / 30 % RH | 85 °C /
85 % RH, water resistance test by dipping of the sensor to water. The
results of the measurements are presented in Figs. 2.10 and 2.11.

From the humidity dependence of impedance of the reference humidity
sensor CT1 (see Fig. 2.6) is clear, that the carbon nanolayer by itself
(without NiPcSNa) cannot be used as a humidity sensor because the

impedance is not sensitive to humidity in the range from 20 to 60 % RH. When NiPcSNa is present the sensors are well sensitive to humidity (Figs. 2.7 – 2.9). The response follows from the dissociation of SO_3Na groups.

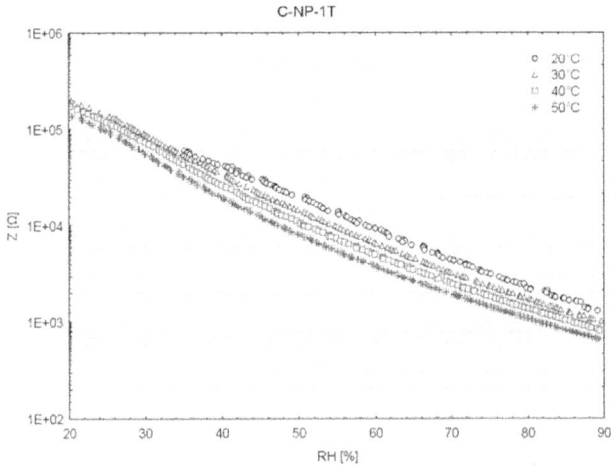

Fig. 2.7. Humidity dependence of impedance of humidity sensor C-NP-1T (C nano layer 6 nm on interdigital electrodes , NiPcS fixed on C surface).

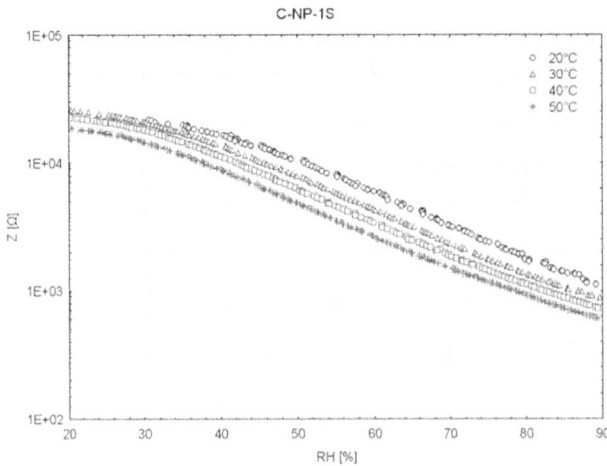

Fig. 2.8. Humidity dependence of impedance of humidity sensor C-NP-1S (C nano layer 11 nm on interdigital electrodes, NiPcSNa fixed on C surface.

38

Chapter 2. Humidity Sensors Based on Substituted Phthalocyanines and either TiO₂
or Carbon Black Nanolayers Modified with them and on the Sandwich Double Layer
Structures, Consisting of Hydrophilic and Hydrophobic Phthalocyanine Derivatives

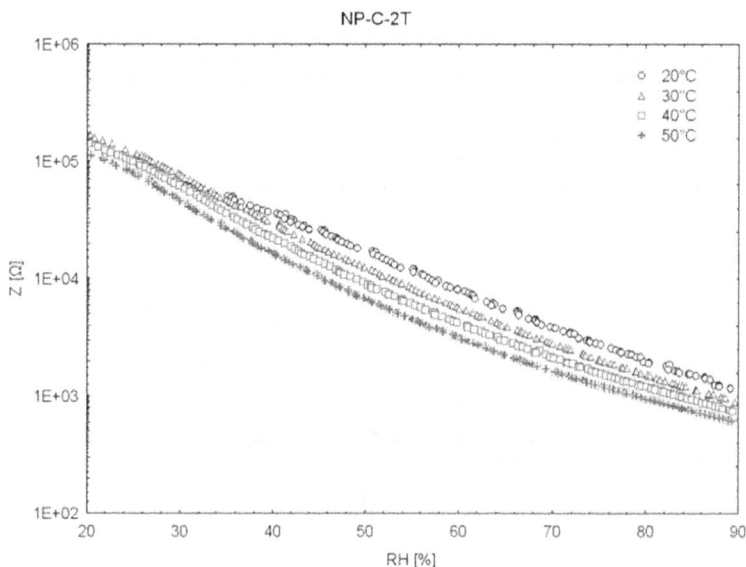

Fig. 2.9. Humidity dependence of impedance of humidity sensor NP-C-2T (NiPc layer on interdigital electrodes, C nanolayer 6 nm fixed on the surface of NiPcSNa layer).

A tendency to saturation at low relative humidity was also observed on sample C-NP-1S, composed of 11 nm C nano layer deposited on interdigital electrodes and NiPcSNa layer fixed on the surface of the C nanolayer (see Fig. 2.8). The low-humidity saturation limits the use of this sensor in technical practice. The saturation is very probably influenced by low conductivity of carbon nanolayer at low humidity. The saturation effect is limited if carbon layer is thinner, cf. Fig. 2.7 (humidity sensor C-NP-1T, composed of carbon nanolayer of 6 nm deposited directly on interdigital electrodes with NiPcSNa fixed on its surface). Similar impedance vs. humidity dependence was observed in the case when NiPcS was directly deposited on the electrode system and C nanolayer 6 nm is fixed on the surface of NiPcSNa layer - sensor NP-C-2T (see Fig. 2.9). Thus, the sensors of the type C-NP-1T and NP-C-2T are recommended for further development. Besides, the humidity sensor C-NP-1T shows good long time stability (see Fig. 2.10) and stability to direct water contact (see Fig. 2.8). Its stability to water was three times higher than that of NiPcS layer directly deposited on Au interdigital electrodes.

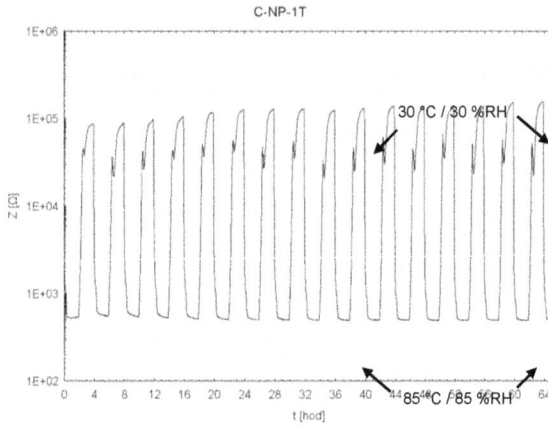

Fig. 2.10. Long-time stability and reversibility tests of impedance of humidity sensor C-NP-1T (C nanolayer 6 nm on interdigital electrodes, NiPcSNa fixed on C surface).

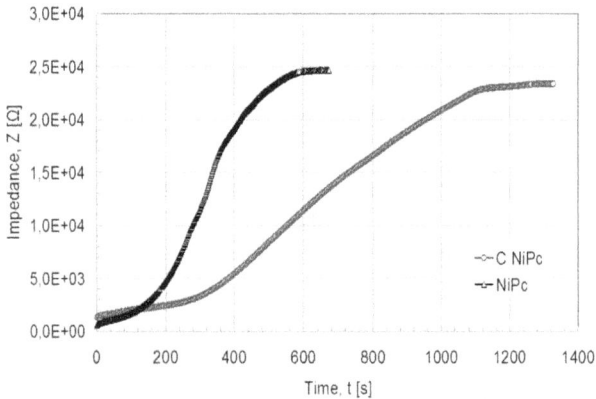

Fig. 2.11. Time dependence of impedance of humidity sensor dipped to the water. NiPc (only NiPcSNa layer without C nanolayer on the surface of interdigital electrodes,), CNiPc (C nanolayer of 6 nm on the surface of interdigital electrodes with NiPcSNa on the C surface).

2.3.3. Results and Discussion

From the humidity dependence of impedance of the reference humidity sensor CT1 (see Fig. 2.6) is clear, that the carbon nanolayer by itself (without NiPcSNa) cannot be used as a humidity sensor because the

Chapter 2. Humidity Sensors Based on Substituted Phthalocyanines and either TiO₂
or Carbon Black Nanolayers Modified with them and on the Sandwich Double Layer
Structures, Consisting of Hydrophilic and Hydrophobic Phthalocyanine Derivatives

impedance is not sensitive to humidity in the range from 20 to 60 % RH. When NiPcSNa is present the sensors are well sensitive to humidity, cf. Figs. 2.7 – 2.9. The response follows from the dissociation of SO_3Na groups.

A tendency to saturation at low relative humidity was also observed on sample C-NP-1S, composed of 11 nm C nano layer deposited on interdigital electrodes and NiPcSNa layer fixed on the surface of the C nanolayer (see Fig. 2.8). The low-humidity saturation limits the use of this sensor in technical practice. The saturation is very probably influenced by low conductivity of carbon nanolayer at low humidity. The saturation effect is limited if carbon layer is thinner (Fig. 2.7) (humidity sensor C-NP-1T, composed of carbon nanolayer of 6 nm deposited directly on interdigital electrodes with NiPcSNa fixed on its surface). Similar impedance vs. humidity dependence was observed in the case when NiPcSNa was directly deposited on the electrode system and C nanolayer 6 nm is fixed on the surface of the NiPcS layer - sensor NP-C-2T (see Fig. 2.9). Thus, the sensors of the type C-NP-1T and NP-C-2T are recommended for further development. Besides, the humidity sensor C-NP-1T shows good long time stability (see Fig. 2.10) and stability to direct water contact (see Fig. 2.11). Its stability when contacted directly to water was three times higher than that of NiPcSNa layer directly deposited on Au interdigital electrodes.

2.4. Humidity Sandwich Sensors Based on Double Layer Structure consisting of Hydrophilic and Hydrophobic Nickel Phthalocyanine Derivatives

2.4.1. Materials

Nickel phthalocyanine ((NiPc) in further text), sodium salt of sulfonated nickel phthalocyanine $NiPc(SO_3Na)_{1-3}$ ((NiPcSNa) in further text), and sulfonamidic nickel phthalocyanine, $NiPc(SO_2NHCH_2CH_2CH_2N(C_2H_5)_2)_{2,8}$ (NiPc(SA) in further text), were synthesized in the way described below. Their chemical structures are shown in Figs. 2.12 and 2.13. The phthalocyanine molecules, mentioned in this work, contained two (x = 2) and three (x = 3) substituents.

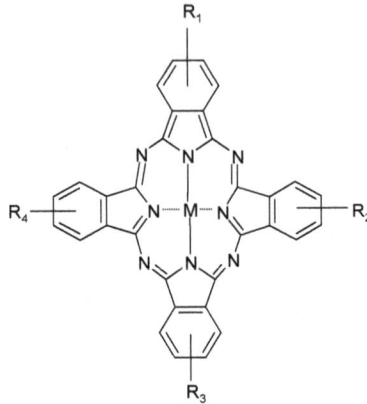

NiPc: M = Ni, R_1, R_2, R_3, R_4 = H

NiPc(SO$_3$Na)$_{1-3}$ (NiPcSNa): R_1, R_2, R_3, R_4 = H or SO$_3$Na

NiPc(SO$_2$NHCH$_2$CH$_2$CH$_2$N(C$_2$H$_5$)$_2$)$_{2,8}$ (NiPc(SA)):

                                              R_1, R_2, R_3, R_4 = H or

                                              SO$_2$NHCH$_2$CH$_2$CH$_2$N(C$_2$H$_5$)$_2$)

Fig. 2.12. Chemical structures of nickel phthalocyanine (NiPc), sodium salt of sulfonated nickel phthalocyanine (NiPcSNa) and nickel phthalocyanine sulfonamide (NiPc(SA)).

2.4.1.1. Sodium Salt of Sulfonated Nickel Phthalocyanine

NiPcS was synthesized in the way as described lower. Its chemical structure is shown in Fig. 2.13. The material used in this work consisted mainly of mono- (x = 1) di- (x = 2) and tri- (x = 3) sulfonated molecules.

Nickel Phthalocyanine (NiPc) was synthesized by the reaction of o-phthalodinitrile with nickel dinitrate. Sodium methanolate was used as a reaction promotor and ammonium molybdenate as a catalyst. The process was carried out in amylalcohol at 138 °C according to Wöhrle at all. [10].

Sulfonated nickel phthalocyanine (NiPcS) was synthesized from NiPc. The sulfonation process was carried out in a four necked glass flask, which was heated in an oil bath, equipped with reflux cooler, agitator, thermometer and dosing funnel. 200 g of fuming sulfuric acid (5 %) was charged into the flask and subsequently 20 g of NiPc was gradually charged through the dosing funnel into the agitated fuming sulfuric acid. The mixture was agitated at 30 °C until all NiPc was fully dissolved.

Chapter 2. Humidity Sensors Based on Substituted Phthalocyanines and either TiO₂
or Carbon Black Nanolayers Modified with them and on the Sandwich Double Layer
Structures, Consisting of Hydrophilic and Hydrophobic Phthalocyanine Derivatives

After that the dosing funnel was removed and the inlet of nitrogen was installed instead of it. The process of sulfonation was then carried out under the nitrogen blanket. The reaction mixture was agitated and heated 30 min. up to temperature 115 °C and subsequently kept for 60 minutes at the adjusted temperature.

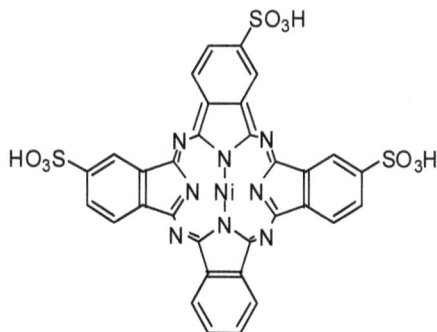

Fig. 2.13. Sulfonated Nickel phthalocyanine, $NiPc(SO_3H)_x$ x = 1 – 3.

After the finishing of the sulfonation process, the reaction mixture was slowly cooled down to laboratory temperature and subsequently charged with a dropping funnel into the vigorously agitated mixture consisting of 2000 g of ice and 1000 g of water. The water suspension was then filtered with a Büchner funnel, the filter cake was washed with distilled water until no sulfate anions were detectable in the filtrate. Thoroughly washed filter cake of NiPcS was dispersed in distilled water and pH was adjusted to pH 11 by using 10 % water solution of NaOH. In this way a sodium salt of sulfonated Nickel phthalocyanine (NiPcSNa) in further text) was prepared. The solution obtained in this way was dried in rotary laboratory vacuum evaporator. Dry product, NiPcSNa, contained 5.9 % mono, 63.9 % di, 29.6 % tri sulfonated species, according to HPLC analysis.

2.4.1.2. Nickel Phthalocyanine Sulfonamide NiPc(SA)

NiPc(SA) was prepared from NiPc by two steps reaction according to Ciba Geigy Corporation Patent [11] (instead of CuPc and N-dimethylaminopropylamine, NiPc and N-diethylaminopropylamine were used as starting materials). In the first step NiPc sulfochloride, containing 11.2 % of sulphur S and 11.5 % of hydrolyzable Cl, was prepared. It represent 2.8 SO$_2$Cl groups and 0.2 SO$_3$H group per NiPc sulfochloride molecule. Subsequently, NiPcSA was prepared from the NiPc sulfochloride by the reaction with N-diethylaminopropylamine in water medium. After the finishing of the reaction the product was filtered off with the Buchner funnel. The filter cake was then washed with distilled water until no traces of N-diethylaminopropylamine were detectable in the filtrate and dried at the temperature of 105 °C until the constant weight. The average molecular constitution of the product was NiPc(SO$_2$NHCH$_2$CH$_2$CH$_2$N(C$_2$H$_5$)$_2$)$_{2.8}$.

2.4.2. Sensor Preparation and Response Detection

3 % water solution of NiPcSNa was deposited by spin coating method on alumina substrate with interdigital system of gold electrodes. After the deposition the system was thermally treated at 80 °C on air. The sensor prepared in this way was stable and responses were reproducible. To improve the stability at high humidity (up to 90 % RH) and temperatures up to 50 °C the hydrophilic NiPcSNa film was covered with hydrophobic NiPc(SA) film. This film was also prepared by spin coating of 3 % chloroform NiPc(SA) solution. After the deposition the sensor was once more thermally treated at 80 °C on air. The sensor prepared in this way was water resistant, stable and responses were reproducible. With increasing humidity the electrical conductivity of the sensor increases (impedance decreases). These parameters were used as output data of the sensor during the humidity influence. The samples were placed in an oven where both temperature and humidity of air could be changed. The sample impedance was measured by the method of capacitor charging though the sample resistance.

2.4.3. Results and Discussion

The dependences of the impedances on relative humidity at various temperatures of the sensor consisting of NiPcSNa film are shown in Fig. 2.14. The impedance decreased with relative humidity (RH). The fall was about two and half orders of magnitude in the humidity range 20 – 90 % RH. Temperature dependence of the impedance was weak. The response of the sensor with humidity was faster for thinner films. The films several tenths of nanometers thick appeared suitable for practical applications. The system was long-time stable up to 90 % RH. The repeated dependences of the impedances on RH are shown in Fig. 2.15. However, when the sensor was kept on air at 80 °C and 90 %

44

Chapter 2. Humidity Sensors Based on Substituted Phthalocyanines and either TiO₂
or Carbon Black Nanolayers Modified with them and on the Sandwich Double Layer
Structures, Consisting of Hydrophilic and Hydrophobic Phthalocyanine Derivatives

RH for a long time or even was covered by water film, the sensor material (NiPcSNa) dissolved and the sensitivity of the device dropped down. To improve the sensor stability at high humidity, two-layer system, consisting of the hydrophilic layer (NiPcSNa) covered by the hydrophobic NiPc(SA) film, was prepared.

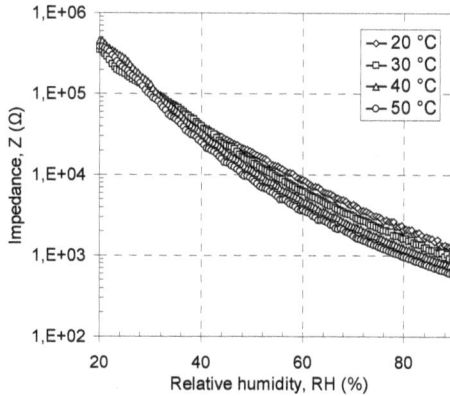

Fig. 2.14. Dependences of the impedances Z vs. relative humidity (RH) of the sensor based on $NiPc(SO_3Na)_{1-4}$ at various temperatures.

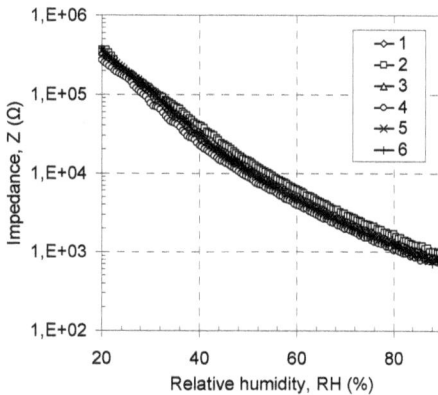

Fig. 2.15. Plots of the impedances Z vs. relative humidity RH of the sensor based on $NiPc(SO_3Na)_{1-4}$ during repeated measurements.

The responses of the impedances on humidity and temperature are shown in Fig. 2.16. The dependences are similar as those of the single layer

sensor but the temperature dependence of the impedance is stronger. The
stability of the sensor improved, the sensor was stable even when it was
put into water, as it follows from Fig. 2.17. The meaning of the figure is
as follows.

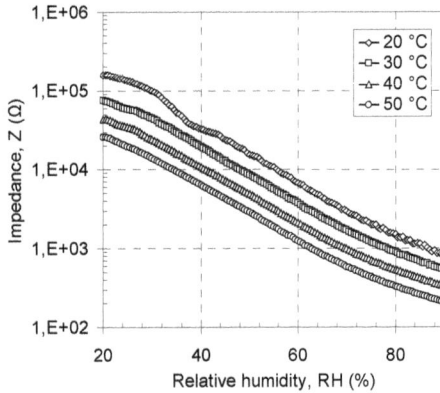

Fig. 2.16. Dependence of the impedance Z vs. relative humidity RH
of the bilayer sensor based on NiPcSNa film covered by
NiPc(SO$_2$NHCH$_2$CH$_2$CH$_2$N(C$_2$H$_5$)$_2$)$_{2.8}$ layer.

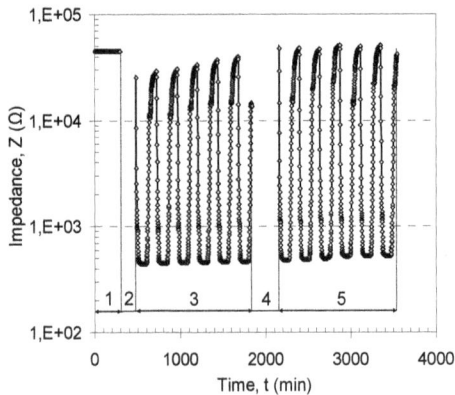

Fig. 2.17. Dependence of the impedance Z on relative humidity RH
of the double layer sensor based on NiPcS$_{1-3}$ layer covered
with NiPc(SO$_2$NHCH$_2$CH$_2$CH$_2$N(C$_2$H$_5$)$_2$)$_{2.8}$ (climatic resistivity test –
details see in the text).

Chapter 2. Humidity Sensors Based on Substituted Phthalocyanines and either TiO₂
or Carbon Black Nanolayers Modified with them and on the Sandwich Double Layer
Structures, Consisting of Hydrophilic and Hydrophobic Phthalocyanine Derivatives

The sensor was first kept under the "normal conditions", 30 °C / 30 % RH and impedance was measured (see segment 1). After that the sensor was put into water (segment 2) for 100 min.

Then, 30 °C / 30 % RH → 85 °C / 90 % RH → 30 °C / 30% RH cycles were realized (segment 3). The sensor was put into water again for 200 min (segment 4) and 30 °C / 30% RH → 85 °C / 90 % RH → 30 °C / 30% RH cycling was repeated (segment 5). The stability of the sensor is evident (stability is determined by the "low parts" of the kinetics curves). The sensitivity is reflected in the "upper parts" of the kinetic curves.

2.5. Conclusions

TiO₂ represents a good material for hybrid humidity sensors. The sensitivity can be improved by the increase of surface area using nanoparticles. However, these humidity sensors are not stable in time. This disadvantage was diminished by the AlPcSNa adsorption. TiO₂ nano layers, prepared from the Solaronix paste, modified, with adsorbed AlPcSNa show good stability and sensitivity to humidity in the range from 20 to 80 % RH. The long-time stability is very good. The response is linear in semilogarithmic plot – logarithm of detection frequency vs. RH and fully reversible which introduces a good background for the humidity sensors fabrication.

Hybrid double layer humidity sensors based on carbon nanolayer and the layer of sodium salt of sulfonated Nickel phthalocyanine (NiPcSNa) show good response to humidity in the range from 20 to 90 % RH. Humidity sensor prepared from carbon nanolayer without NiPcSNa was not sensitive to water in the range from 20 to 60 % RH.

C-NP-1S sensor, composed of 11 nm C nanolayer deposited on interdigital electrodes and NiPcSNa layer fixed on it, shows response saturation at low humidity which limits its using in technical practice. Sensor C-NP-1T, composed of 6 nm carbon nanolayer deposited on interdigital electrodes and NiPcSNa layer fixed on it shows good response to humidity and is recommended for further development. Similarly behaves sensor NP-C-2T composed of NiPcSNa layer deposited on the interdigital electrodes and 6 nm thick carbon nanolayer fixed on it.

The sandwich double layer humidity sensor, based on NiPcSNa was improved. To improve the stability of the sensor at high temperature and humidity the second hydrophobic NiPc(SA) layer was applied over the first NiPcSNa hydrophilic layer. The "double layer," sandwich sensor shows very good reversibility and stability even under the condition 85 °C / 90 % RH. Even a direct contact with water 200 min. long does not influence the sensor performance. The sensors prepared from thinner films are faster. The films of several tenths of nanometers thick seems to be suitable for practical applications.

Acknowledgements

This research has been supported by the European Regional Development Fund and the Ministry of Education, Youth and Sports of the Czech Republic under the Regional Innovation Centre for Electrical Engineering (RICE), project No. CZ.1.05/2.1.00/03.0094.

References

[1]. Arai, H.; Seiyama, T. Humidity Sensors, in Sensors: a comprehensive survey, W. Gopel, J. Hesse and J. N. Zemel (Eds.), *VCH, Weinheim*, Vol. 3, 1992, pp. 981-1012.

[2]. N. Yamazoe, N., Shimizu, Y. Humidity sensors: Principles and applications, *Sensors and Actuators B, Vol.* 10, Issue 3-4, 12 November 1986, p. 379-398.

[3]. Traversa, E., et al., Electrical humidity response of sol-gel processed undoped and alkali doped TiO_2–Al_2O_3 thin films, *Journal of the European Ceramic Society*, Vol. 19, Issues 6–7, June 1999, pp. 753–758.

[4]. Bearzotti, I., et al., Highly ethynylated polymers: synthesis and applications for humidity sensors, *Sensor and Actuators B,* Volume 76, Issues 1–3, 1 June 2001, pp. 316–321.

[5]. Sakai, Y. Sadaoka, Y., Matsuguchi, M. Humidity sensors based on polymer thin films, *Sensor and Actuators B,* Issues 1–3, September 1996, pp. 85–90.

[6]. Rakusan J., Karaskova M., Hamacek A., Reboun J., Rais D., Nespurek S., Humidity sensor based on TiO_2 nanoparticles and sodium salt of sulfonated AlOH phthalocyanine, in *Proceedings of International Conference NANOCON'10,* Olomouc, Czech Republic, October 12th –14th 2010, pp. 177 – 181.

[7]. Rakusan J., Karaskova M., Hamacek A., Reboun J., Stulik J, Nespurek Sb, and Kubac L.: Phthalocyanine hybrid humidity sensors based on carbon nanoparticles and substituted phthalocyanines, in *Proceedings of*

International Conference NANOCON'12, Brno, Czech Republic, October $23^{rd} - 25^{th}$ 2012, pp. 416 – 420.

[8]. Rakusan J., Karaskova M., Hamacek A., Reboun J., Kubac L., Cerny J., Rais D., and Nespurek S.: Double layer humidity sensor based on phthalocyanine derivatives, in *Proceedings of International Conference NANOCO'11*, Brno, Czech Republic, September $20^{th} - 23^{rd}$ 2011, pp. 1241 – 1247.

[9]. Rakusan J., Karaskova M., Cerny J., Kubac L., Hamacek A., Reboun J., Czech Patent No. 303823, 2013.

[10]. Wöhrle D., Schnurpfeil G, Knothe G., Efficient Synthesis of Phthalocyanines and Related Macrocyclic Compounds in the Presence of Organic Bases, *Dyes and Pigments,* 18, 2, 1992, pp. 91-102.

[11]. Ciba-Geigy Corporation, *US Patent No. 4,318,883,* 1982.

Chapter 3

Fault Tolerance and Fault Management Issues in Wireless Sensor Networks: Recent Developments

Radosveta Sokullu, Özlem Karaca Akkan

3.1. Definition of Fault Management and Fault Tolerance in WSN

Fault tolerance (FT) is a function extremely important for the operation of WSN, defined as "the ability of the network to continue functioning in the presence of link and/or node failures" [1]. The level of fault tolerance is highly dependent on the specific WSN application. According to the authors of [2] a good fault management (FM) solution providing network fault tolerance has to comply with two very important characteristics: it has to be both energy-aware and vulnerability aware. The authors of [3] present a detailed investigation on the different aspects of providing fault management in WSN. According to their taxonomy, functions realizing fault management can be summarized in four groups:

Fault Prevention: functions that allow preventing and minimizing the occurrence of faults;

a) Ensuring full network coverage and connectivity during design and development stage;

b) Providing constant monitoring of the network and the individual nodes and triggering reactive actions when required;

c) Providing redundancy in nodes and their connections.

Fault Detection: functions that allow detecting abnormal network operation signs based on monitoring specific networks performance parameters like packet loss, delay etc.

Fault Identification: using the information collected from monitoring the network operation different hypotheses can be developed for the possible origin and type of the faults.

Fault Recovery: functions that provide minimizing the effect of the faults or if possible, provide recovery.

Faults in WSN might be due to a number of reasons from the continuously changing wireless channel, to physical damage and disruption. Besides, a lot of the faults might be due to reasons very different from those in traditional networks. For example while in wired networks, packet loss or extensive delay is a clear indication of congestion, in WSN we might define at least three different causes leading to this result: node response failure (due to battery depletion, node failure or mobility), link failure or congestion in the network. Another important factor to be considered is that WSN are very limited in resources – both power and processing resources – and traditional approaches for dealing with network faults are generally not applicable. Thus, the fault management framework (FMF) is a complete system that takes into consideration the different aspects of fault tolerant operation in WSN. It is defined as a generic unified structure that takes into consideration diverse sources of faults as well as application specific operation of WSN. [4].

There is a sizeable amount of literature dealing with the subject of fault tolerance in WSN. Different authors have suggested quite diverse methods and provided solutions including one or more of the major functions required for dealing with faults. Similarly there exist a large number of papers summarizing specific fault recovery methods, fault tolerant routing, and fault tolerant protocols. The focus of this chapter however, is on complete systems, providing all or most of the required operations for Fault Management, that is why a brief overview of surveys related to Fault Management Frameworks (FMF) will be included, followed by a more detailed description of some representative frameworks.

The work presented in [3] is an earlier work that provides an extensive summary of the existing method articles grouping them based on the aspect or aspects of fault tolerance they are addressing: fault prevention, fault detection, fault identification and fault recovery. The more recent study [4] includes a brief overview of existing frameworks from another perspective - reviewing the major approaches and network parameters utilized for dealing with faults, namely remaining energy of the node, buffer level, packet loss, latency and network lifetime.

In their survey [5], M. Yu et al. divide the fault management process into three phases - fault detection, diagnosis and recovery. The authors classify the existing frameworks into centralized, distributed and hierarchical models.

The centralized model is represented by a central controller, which is responsible for fault maintenance of the overall network. In order to construct the global view of the network, the central controller keeps updating nodes' states by message exchanges and aggregates the data into information models, in terms of specific metrics [6], topology models [7], topology and energy maps [8], WSN models [9], and cluster topology models [10]. M. Yu et al. point out that centralized architecture does not scale well and will lose its efficiency and effectiveness when the network expands. Some worth noting examples of centralized FTF are MANNA [11], Symphony [12] and WinMS [13].

MANNA, a centralized, comprehensive management framework, is one of the earliest suggested FTF. In MANNA, each sensor node is assigned a role as manager or agent, and the manager builds coverage and energy models based on the information received from the agents (sensor nodes), which are later used to perform fault detection and identification. MANNA is considered a basis for centralized fault management frameworks. Later on M. Yu et al. proposed an improved architecture "Sympathy", in which nodes are classified as "Sympathy-sink" or "Sympathy-node".[12] The Sympathy-sink is a non-resource constrained node, responsible for operations related to fault management. However, both MANNA and Sympathy fail to provide fault recovery. A centralized fault management scheme which provides also corrective and preventive actions is WinMS. [13] It is based on a schedule-driven MAC protocol FlexiMAC, which is used to collect and disseminate management data, and provide autonomy for individual sensor nodes to perform management functions. It is an example of a central network management scheme that uses the central manager with a global knowledge of the network to reliably execute corrective and preventive fault management and network maintenance, thus increasing the FT of the whole network.

Distributed frameworks allow for the division of the network into multi-regions, each sub-region consisting of a central manager and a certain number of sensor nodes, where the fault management tasks are assigned well-proportioned to every region in a distributed way. It directly communicates with other regional managers in a coordinated fashion. Yu and his colleagues argue that the distributed design is more energy efficient and enhances system response time.

Some earlier examples, representative of the distributed model frameworks can be found in [14-15], and [16]. The basic idea of distributed model is to have each of the independent sensors make a local decision and forward these decisions to the sink generating a global decision. Clustering is an efficient approach for building scalable and energy-balanced fault management schemes. In such examples intermediate managers are used distribute management functions. However, these intermediate managers do not directly communicate with each other but are rather involved to manage nodes in their subnet and report local decisions to a superior manager, which then is directed to the central manager utility. Such examples can be found [17] and [18]. In the first one a three-layered fault management structure is suggested. The bottom layer consists of sensor nodes which send data to their corresponding cluster head. The intermediate layer comprises the Cluster Heads (CHs), which make decisions about fault events within the individual subnet. The decisions from CHs are further transmitted to the fusion center (Sink) to inform it if there is or isn't an abnormal situation in the specific sub-region. In the second one, the authors propose a hierarchical architecture, where the central manager is at the highest level and placed at the sink node, the intermediate manager works at the cluster head level and there are management agents which are the normal sensor nodes. The intermediate managers are used to receive fault information from the in-cluster sensor nodes. However, they do not communicate with each other and work independently. The central network manager has the global knowledge of the network states and gathers the global knowledge from the underlying network layers and intermediate network managers to achieve fault detection and fault recovery of the cluster heads.

Another more recent example of a distributed fault management system designed to overcome the single-point-of-failure problem is discussed in [19]. In a lot of WSN applications, there is a single sink gathering the data taken by the sensors which can be placed anywhere in the field. When sensors are not connected directly to the sink, they must send data in a multi-hop fashion, by hopping through other sensors. To make the operation more time efficient the sink is usually placed at the center of the process to effectively analyze the measured data and make the final decision. This however, leads to single-point-of-failure problem which drastically reduces the fault tolerance of the whole network. In order to eliminate the problem, without incurring additional delays, the authors propose a distributed scheme to enhance the accuracy of detecting the result in an autonomous local sensor network without the need of a centralized sink. Generally, to derive the final result in WSNs a consensus algorithm is required. However, when sensors in the network are disabled due to the physical damage or lack of power reaching a consensus is a problem. Thus, it is still an open research issue to ensure joint decision-making among sensors in the presence of sensor fault and transmission media faults. In the study, the authors have proposed a fault tolerant solution to the consensus problem by exploiting the "Byzantine Agreement" algorithm. The solution allows the sensors to perform a required action even when a limited amount of sensors are faulty and the transmission media is disturbed. According to the suggested protocol the sensors are divided into different autonomous local WSN and the solution is defined for each autonomous area. As a result, all sensors in the system can decide and do corresponding actions without the sink, and thus a) The single-point of failure problem is solved; b) The hopping process time is reduced. The suggested decision mechanism as part of a fault tolerant framework works well for separate sensor or media faults but improvements are required for the combined case when there exist simultaneously a sensor fault and transmission media fault.

Finally, the hierarchical model is a hybrid between the centralized and distributed approach. It uses intermediate managers to distribute manager functions. However, these intermediate managers do not directly communicate with each other. Each manager is responsible for managing nodes in its sub-network and reporting to its high-level central manager. An example of a three-layered hierarchical sensor network structure can be found in [18]. Another example of hierarchical model can be found in [19], where in order to accomplish fault management hierarchical mobile agent-based policy management architecture is suggested. It contains the Policy Manager (PM) at the highest level, Local Policy Agents (LPA) which manage each sensor node, and a Cluster Policy Agent (CPA) at the intermediate level.

Another, more recent survey and classification of FMF can be found in [20]. The authors of that paper, similar to the above mentioned [5], classify the frameworks into centralized, distributed and hierarchical and provide examples of each. Furthermore, they also introduce a comparison of recent frameworks for fault management in WSN based on the following parameters: lightweight, scalability and autonomy. Implementation of any additional process, including fault management, creates an additional burden to the inherently resource limited WSN, that is why lightweight is considered a criteria of utmost importance. On the other hand, WSNs by default consist of numerous nodes, and depending on the application their size might vary from tens to several thousands, thus scalability is always an important issue to be considered. Autonomy on the other hand refers to the ability of the fault management framework to provide capability of self-organization to adapt to dynamics of network.

Another very interesting survey can be found in [21]. The authors start with a detailed definition of the possible types of faults is WSN and discuss the effects they have on the operation of the WSN. They also provide a brief evaluation of some FM frameworks based on functionality (fault monitoring and fault detection), architecture, and major parameters like energy efficiency, scalability and their response to different design issues. MANNA, is classified as proactive, hierarchical, adaptive and scalable fault management system. It continuously monitors the WSN and keeps on updating the management information base, which however is a very dynamic and energy exhausting operation. BOSS [8] is classified as proactive, centralized, energy efficient, adaptive and scalable management system which is end-user dependent, since the end-user is responsible for observing and taking actions. CRAFT [22] is classified as reactive, hierarchical, energy efficient and however not self-configurable. It has a high maintenance value, since it relies on periodical data checking using checkpoints. RS (Ride Sharing) [23] is defined as reactive, hierarchical, energy efficient, adaptive and scalable fault detection scheme which is based on redundancy. However this incurs extra overhead and increased overhearing by the nodes which leads to node buffer overflows.

Following the brief overview of existing surveys and classifications of fault management frameworks, in the next section we describe in more detail some recently suggested solutions for providing fault tolerance, which have not been covered in the above mentioned literature. More specifically we discuss 4 very recently published works presenting an Energy Aware Fault Tolerant Framework, a Generic Component based one, a Relay based one and a Cross Layer Design based Fault Tolerance Management Module.

3.2. Fault Tolerance Management Frameworks

3.2.1. Energy Aware Framework

An Energy Aware Fault Tolerant Framework for WSNs is proposed in [24]. The framework comprises three fundamental layers: Fault

56

Detection layer, Fault Diagnosis layer and Fault Recovery layer. The Fault Detection layer including components such as Sensor Monitor Listener, Sensor History Manager, Monitor History DB, Sensor Fault Detector and Sensor Fault Predictors is responsible for detecting any fault occurrence and predicting faults. Sensor Monitor Listener checks each sensor node or network regularly. Monitor History DB is a repository of historical events. Sensor History Manager runs the relevant query to match the information with the stored data. While matching the query Sensor History Manager can also send the information to Sensor Fault Detector and Sensor Fault Predictor simultaneously. In case, the sensed data does not match with any of the existing one then the sensor fault detector records the event in the database. Detector and/or predictor communicate to fault diagnosis layer having Sensor Node Diagnosis Manager and WSN Diagnosis Manager. Fault can be either of sensor node level or network level, so that this layer has two diagnostic unit. The Sensor Node Diagnosis Manager finds out the actual origin of the fault and its attributes and which hardware in the sensor node is affected. The obtained information is further conveyed to the WSN Diagnosis Manager, who manages to the total diagnosis work for the whole network. Sensor Node Diagnosis Manager and WSN Diagnosis Manager both communicate to Recovery Planner which is located in the Fault Recovery layer. It has four components: Network level recovery Manager, Node level recovery Manager, Recovery Planner and Notification. Recovery Planner contacts Node level recovery Manager and Network level recovery Manager to recover the fault.

3.2.2. Generic Component Framework - FlexFT

In the study [25] the use of a generic component based framework for the construction of adaptive fault tolerant systems that can integrate and re-use technologies and deploy them across heterogeneous devices is discussed. The authors examine two types of heterogeneity: Device heterogeneity and Software language/middleware heterogeneity. Their aim is to define approaches that can enable the development of middleware solutions addressing different programming models in different environments. The suggested tool, FlexFT, is a generic tool for constructing reliable systems that can deal with both hardware and software heterogeneity for providing fault tolerance in WSN. It consists of a minimal policy-free microkernel where fault tolerance policies are implemented as demanded by the specific application or deployment. The framework can be extended by adding component plugins, which can be removed when no longer required. FlexFT supports dynamic adaptation, as the plugins can be reconfigured at runtime. It is intended for constructing reliable systems based on a wide range of software technologies targeted to a variety of hardware platforms. Furthermore, the authors present the implemented framework prototype and evaluation of its potential benefits.

3.2.3. Relay Based Framework

Several research groups, among them [26, 27] and [28] deal with the subject of fault management taking into consideration a very interesting trend that has emerged in recent years: the relay nodes. Relay nodes are special functionality nodes in WSN, proposed to achieve objectives like extending network lifetime, balancing data gathering and reducing required transmission distance thus improving network connectivity and enhancing the network fault tolerance as a whole [29-32]. Such relay nodes can be provisioned with higher power and enhanced capabilities, as compared to the general sensor nodes, and are ideally suited to serve as cluster heads in a hierarchical, two-tier sensor network. [26, 33-38]. In such networks, sensor nodes are partitioned into clusters and a relay node, which acts as a cluster head in each cluster. Thus by introducing the relay nodes as a specific type of redundancy in WSN, [27] address the problem of fault tolerance from a different angle. It is well known that, given a certain WSN deployment, finding the exact locations to place a minimum number of relay nodes such that, each sensor node is covered by at least one relay node, is a computationally difficult problem. In addition, for successful and reliable data communication, the network with relay nodes needs to be connected, as well as resilient to node failures. So, in their work the authors propose a novel integrated Integer Linear Program (ILP) formulation and two heuristic algorithms, which, not only define a suitable placement strategy for the relay nodes, but also assign the sensor nodes to the clusters and determine a load-balanced routing scheme. As a result, within the suggested framework the desired levels of fault tolerance both for the sensor nodes and the relay nodes can be secured, and the predefined performance guarantees are met with respect to network lifetime.

3.2.4. Cross Layer Design Based Framework - FTMM

In [4] a Fault Tolerance Management Module (FTMM) is defined as a framework suitable to the requirements, limitations and specifics of

WSN. It encompasses methods for fault detection, fault prevention, fault management and recovery. The suggested solution is based on the Cross Layer Design (CLD) approach, which has proved to be an important tool in increasing network performance for WSNs. When there exist severe faults in WSN, MAC and routing protocols must provide new links and routes, transport protocols must adaptively resolve how to retransmit, application layer protocols and must decide what level of loss is tolerable. Thus, multiple levels of redundancy may be needed and a cross layer approach exploring the interactions among different layers is desirable as also pointed out in [3]. Therefore, FTMM is a distributed system and build using a CLD methodology, which allows a unified approach to fault detection, identification and recovery. Conceptually it is based on previous work presented in [39] and [40]. The fault management system has a modular structure, with major modular blocks providing fault prevention procedures, fault detection and identification procedures and fault recovery. Nodes act in an independent manner based on their own state information and information collected from one-hop neighboring nodes. Furthermore all the procedures for fault management rely on the cross layer information exchange, both from lower to upper layers and from upper to lower protocol layers. Because of decentralized decision and operation mechanism the size of the network can be increased with very little extra control message traffic. Furthermore, using the CLD approach and considering the functions of all layers of the protocol stack in a unified manner allows for optimization and simplification. FTMM does not assign different roles and role distribution among the nodes, so additive control packet traffic is kept to a minimum.

The proposed fault management module operational structure is shown in Fig. 3.1.

3.2.4.1. Major Building Blocks of FTMM

The network operates in a fully distributed manner. Each node decides by itself whether to participate in the communication process or not, based on the value of the "participation determination" (PD) parameter. The value of PD is determined by the energy level of the node at that moment, by the buffer occupancy, the number of packets sent, the SNR (inferred from the received packet) as well as the response coming from the FTM module. The PD is set to 1 if the first four conditions in Eq. (3.1) are satisfied. Each condition in Eq. (3.1) constitutes certain communication functionality. PD also checks the value coming from FTMM and applies different threshold levels for each parameter considered:

$$PD = \begin{cases} 1, \text{if} \begin{cases} \xi_{RTS} \geq \xi_{RTS}^{thrv} \\ P_i^{TX} \leq k.P_{TX}^{thrv} \\ B \leq B_{max}^{thrv} \\ E_{rem} \geq E_{rem}^{min\,thrv} \\ FT_{action} \leq FT_{action}^{thrv} \end{cases} \\ 0, \text{otherwise} \end{cases} \tag{3.1}$$

where ξ_{RTS} is the current value of the SNR obtained from the received RTS packet, ξ_{RTS}^{thrv} is the threshold value of accepted min SNR; P_i^{TX} is the number of packets transmitted by node i and the P_{TX}^{thrv} is the threshold accepted as a corresponding maximum; B and B_{max}^{thrv} are respectively the current value of the buffer and the max allowed buffer occupation threshold value per node; E_{rem}, and $E_{rem}^{min\,thrv}$ are the currently remaining energy level and the accepted threshold value; FT_{action} and FT_{action}^{thrv} are respectively the values coming from the FTMM module and the threshold, which corresponds to the set of initial system values.

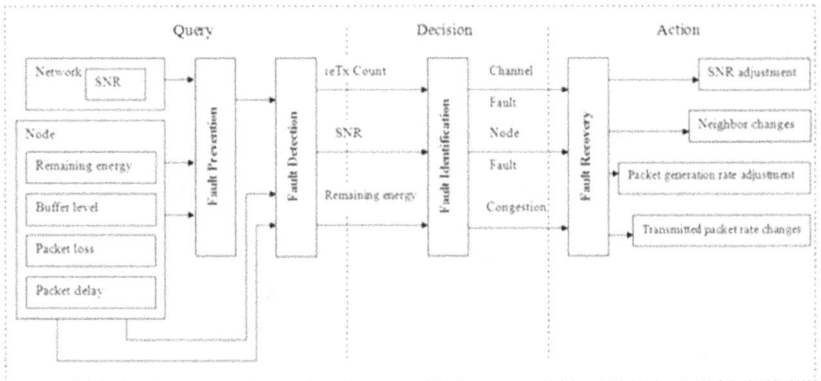

Fig. 3.1. Proposed Fault Management module operational structure.

If necessary, FTMM module changes the FT_{action} threshold values in calculating the PD parameter in such a way that will help mitigate the effect of faults. If FT_{action} is smaller or equal to a specific threshold value, the process is considered a "regular process", if not the FTMM reduces the threshold values incrementally using the FT_{action} parameter trying to alleviate the effect of the faults experienced at a specific node on the operation of the network as a whole. (see [41] for details). When network conditions return to normal, the FTMM module increases the FT_{action} parameter values incrementally to achieve better network efficiency. FTMM operation will be explained in following sub-sections in more detail.

$$
\begin{aligned}
&\text{if } FT_{action}{}^{+} \quad
\begin{cases}
k_{t1} = k_{t0} + k_{level} \\
P_{ii_{t1}} = P_{ii_{t0}} + \beta
\end{cases} \\[2em]
&\text{if } FT_{action}{}^{-} \quad
\begin{cases}
k_{t1} = k_{t0} - k_{level} \\
P_{ii_{t1}} = \dfrac{P_{ii_{t0}}}{\alpha}
\end{cases}
\end{aligned}
\tag{3.2}
$$

In order to understand the operation of the FTMM module better, first the sub-modules comprising FTMM and their algorithms are explained in more detail below.

Sub-module: Fault Prevention

Because of the most restricted source in WSN is nodes' energy, remaining energy level of nodes are very important early-indication for possible errors. If a node's remaining energy is critically reduced, then the node cannot successfully transmit its own packets or relay its neighbors' ones. During the communication, intermediate nodes in the network relay neighbors' packets. In case of congestion the sending/receiving buffer of a node may be critically full. Therefore buffer occupancy may be another indicator that gives information about network conditions. If buffer of intermediate node is critically full, then continuing to participate in the communication with its neighbors might only increase packet loss and latency, diminishing the general performance of the network. Thus, a node, monitoring its remaining energy and buffer space, can decline to participate or withdraw from the communication process based on the value of PD parameter. This solution reduces the possibility of error occurrence and untimely interruption of the network operation. Furthermore, thresholds for these values are introduced, which allow the node not only to withdraw but also to limit its participation without fully withdrawing from it. For example, if the buffer level and/or the remaining energy are still above (respectively below) a critical level but getting close to the predefined threshold, the node has the option to reduce the number of transmitted packets, giving priority to the ones it generates and declining to relay others' packets. Thus, by the introduction of thresholds, the withdrawal of the communication process might be achieved in a graceful manner without abruptly interrupting the network operation and providing possibilities for fault prevention.

Sub-module: Fault Detection

For fault detection it is very important to monitor the state of the network. In the suggested system this is done in an indirect distributed manner. None of the nodes has a complete picture of the whole network, however based on continuous monitoring of its own node parameters (buffer level, remaining energy - as well as that of the direct neighbors - and the retransmission count – reTX) and the channel parameters (signal-to-noise ratio) it can create quite a useful picture of its close neighborhood. When a node comes out of sleep state it broadcasts a short "state message" that is used to refresh the information in the so called "neighbors lists" of its one-hop neighbors. It also carries the current energy value for that specific node. Reduced buffer levels as well as indications of low SNR are an early warning for the possibility of errors and increased packet loss and latency come as a confirmation. Signal-to-noise ratio (SNR) of the received packet can be indicator of faults originating due to worsening channel conditions and possibilities for reduced network capacity. Thus SNR is also monitored carefully. When the information about the channel parameters and the node parameters is handled together, it is possible to provide an early warning to congestion or reduced performance in the network (Fig. 3.2).

Sub-module: Fault Identification

For fault identification each node tries to make an independent decision about the source of the fault. In Fig. 3.3 we present the main algorithm that runs on each node for the identification of the possible sources of occurring faults. Depending on the reason for the fault the following three major cases have been defined: channel fault, node fault, congestion. Channel faults occur when the SNR of the communication between two nodes gets worse; the retransmission number increases (and also the buffer occupancy level increases); low SNR leads to increased loss of packets and increased latency.

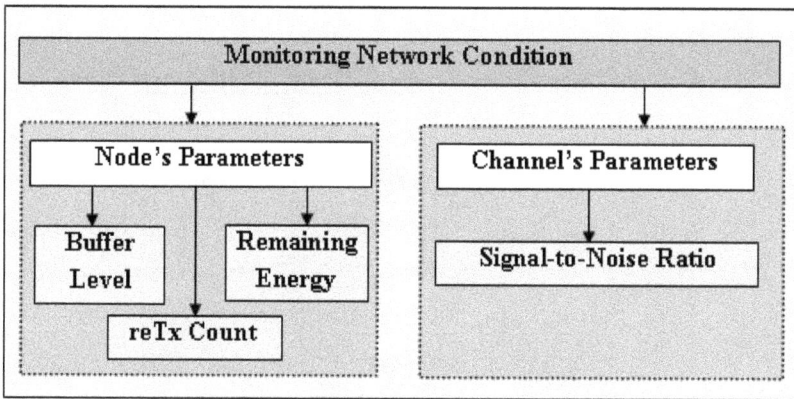

Fig. 3.2. Proposed Fault Detection scheme.

Running the algorithm for fault identification given in Fig. 3.3 and based on the above mentioned information each node can determine that a channel fault has occurred. Node fault is defined as the case for which the SNR of the channel is high but there are disturbances in the packet flow between the node and a specific immediate neighbor, usually indicated by increased packet loss and delay. If the channel quality is within limits, the remaining energy is high but the retransmission rate and the latency are dangerously above expected limits than faults might be identified as due to congestion.

Sub-module: Fault Recovery

Based on the output of the fault identification sub- module which provides information regarding the possible causes of faults the fault recovery module performs its actions (Fig. 3.4). Since according to the distributed system model adopted, each node decides on its own whether or not to participate in the communication process based on the values of the PD parameters, it is possible to adjust the SNR threshold to reflect the worsening channel conditions. Adjusting the SNR threshold allows for continuation of the packet flow thus reducing the packet loss and increasing the number of retransmissions until the channel quality returns to normal. When a node fault is identified, the fault recovery procedure might prompt for a possible change in the immediate neighbors, based on the neighbors' list information.

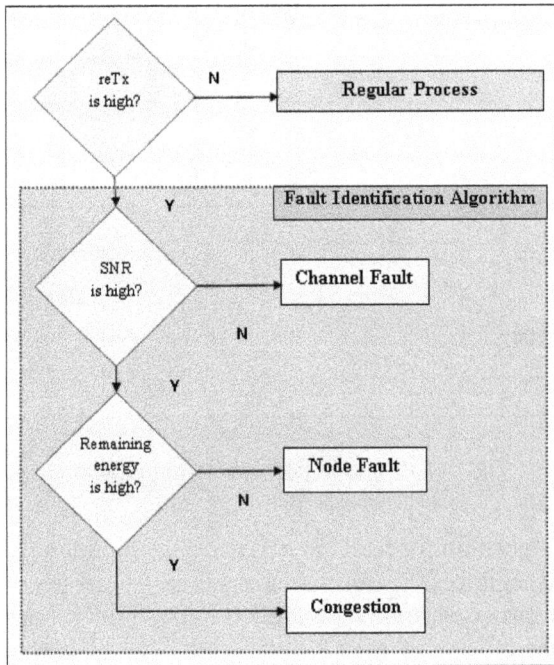

Fig. 3.3. Proposed Fault Identification algorithm.

In this distributed system it is not possible to fight congestion on a network scale. However, if a number of nodes that have identified congestion in a given area reduce their rate of transmission and retransmission for a certain period of time to a plausible minimum this would greatly help and possibly resolve the situation. Decreasing the packet generation rate will result in decreased traffic thus the network is given a chance to recover to its normal course of operation. However it is obvious that this recovery mechanism is at the price of increased delay.

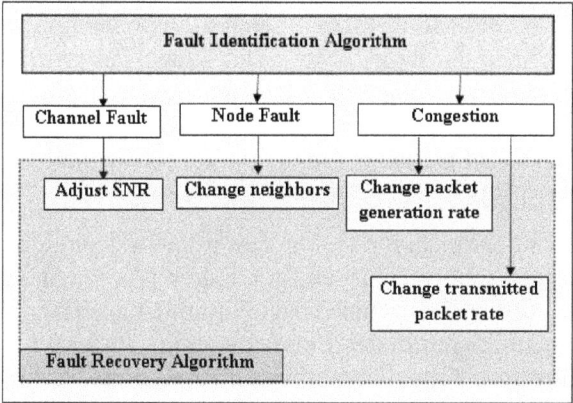

Fig. 3.4. Proposed Fault Recovery algorithm.

3.2.4.2. Fault Tolerance Management Module (FTMM) operation

After defining the major building blocks in this section operation stages of FTMM are described. An overview of the FTMM functional structure was given in Fig. 3.1.

The following stages have been defined: Query, Decision and Action. The Query stage realizes the functions of possible fault prevention and fault identification. During this stage, each node collects information about its own state and the state of the immediate surroundings. Nodes keep track of their remaining energy and buffer levels, as well as the number of retransmitted packets (reTX count) as an early warning for the arise of error. The SNR of the incoming packets is used as an indication of the channel state together with information about the remaining energy level of its immediate neighbors. Based on the knowledge of the buffer state and the reTx count each node can get an estimation of the increase in packet loss. Furthermore, time stamping (relayed packets and RTS messages) allows for the evaluation of the incurred delay in communicating with the neighboring nodes. So, at the Query stage, the node can choose to take fault prevention measures by gracefully withdrawing from further participation in the communication process, and/or can proceed with fault detection based on the acquired information SNR, packet loss and delay.

During the Decision stage, the cause of the error is investigated, based on the fault identification algorithm given in Fig. 3.3. The data collected in the Query stage are used as an input to the Decision stage. The fault

65

identification algorithm is triggered by three major inputs: values of the SNR below a predefined threshold, value of the remaining buffer space below a predefined threshold, or the reTX count. The re-transmit (reTx) count is used in determining whether a transmission error has occurred. The default value for the reTx count in the initial set of node values is set to 3 and a value 4 is a trigger to the fault identification procedure (for details refer to [41]).

After identifying the error the system enters the next stage, Action stage. During this stage actions are taken for eliminating or reducing the effect of the errors (see Fig. 3.4), which lead to limiting the traffic load and thus allow the network to return to normal operation. At the Action stage, the FTMM realizes the following actions: change in packet transmit rate, adjustment of packet generation rate, change of neighbor and change of the SNR level used for packet acceptance at the specific node.

According to the communication model considered, each node transmits two types of packets: generated by the node itself and packets relayed from other neighbor nodes. So, in case of any indication of increased traffic (congestion) the first action that the fault recovery module takes is reducing the rate of the relayed packets, giving priority to the nodes own generated packets. As a result, by responding less to transmit requests from neighbors the node reduces its total transmit rate while preserving the transmit rate for its own packets constant, while the rejected neighbor will detect a faulty neighbor. If the signs persist the node will reduce its packet generation level. If a faulty neighbor node is detected, the node itself will search in its neighbors' list for an exchange.

Finally, during the Action stage the node can also act for adjusting the SNR as explained above. The operation of the FTMM described here relies on normal packet exchange between the nodes and the introduction of very short "state report" control message (4 bytes). Thus, as a whole the FTMM takes advantage of the available information and the cross layer design incurring minimal control overhead.

The FTMM module presented in this subsection relies on fully distributed operation. The performance (in terms of packet loss, delay, and energy consumption) of a WSN enhanced with FTMM has been compared with other cross-layer design solutions for WSNs, which do not include fault management. The simulation results support the claim of the authors that the introduction of a unified approach to fault management improves the network performance in terms of packet loss, delay, and power consumption. Further details on the comparison and simulation results can be found in [4] and [41].

3.3. Conclusion

In this chapter we have discussed fault tolerance and fault management schemes for wireless sensor networks. First a definition of these concepts is provided followed by a summary of existing surveys on fault management in WSNs and classification of Fault Management Frameworks. From there on several interesting and recently published proposals for providing fault tolerance have been discussed in more detail. These include and Energy Aware Fault Tolerant Framework, a Generic Component Framework, a Relay based Framework and a Cross Layer Design based Fault Tolerance Management Module.

References

[1]. I. F. Akyildiz, W. Su, Y. Sankarasubramaniam, E. Cayirci, A Survey on Sensor Networks, *IEEE Communications Magazine*, 40, 8, 2002, pp. 102-114.

[2]. N. M. Khan, A. Ihsan, Z. Khalid, G. Ahmed, R. Ramer, A. A Kavokin, Quasi centralized clustering approach for an energy-efficient and vulnerability-aware routing in wireless sensor networks, in *Proceedings of the 1st ACM International Workshop on Heterogeneous Sensor and Actor Networks (HeterSanet '08)*, Hong Kong, China, 2008, pp. 67-72.

[3]. L. Paradis, Q. Han, A Survey of Fault Management in Wireless Sensor Networks, *Journal of Network and Systems Management*, Vol. 15, No. 2, 2007, pp. 171-190.

[4]. L. O. Karaca, R. Sokullu, A Cross-Layer Fault Tolerance Management Module in Wireless Sensor Networks, *Journal of Zhejiang University – Science C – Computers & Electronics*, 13, 9, 2012.

[5]. M. Yu, H. Mokhtar, M. Merabti, A Survey on Fault Management in Wireless Sensor Networks, in *Proceedings of the 8th Annual Postgraduate Symposium 'The Convergence of Telecommunications, Networking and Broadcasting' (PGNet'07)*, School of Computing & Mathematical Science, J. M. University, Liverpool, 2007.

[6]. N. Ramanathan, K. Chang, R. Kapur, L. Girod, E. Kohler, D. Estrin, Sympathy for the Sensor Network Debugger, in *Proceedings of the 3rd IEEE Conference on Embedded Networked Sensor Systems*, USA, 2005.

[7]. J. Staddon, D. Balfanz, G. Durfee, Efficient Tracing of Failed Nodes in Sensor Networks, in *Proceedings of the 1st ACM International Workshop on Wireless Sensor Networks and Applications*, USA, 2002, pp. 122-130.

[8]. W. L. Lee, A. Datta, R. Cardell-Oliver, Network Management in Wireless Sensor Networks, in Handbook of Mobile Ad Hoc and Pervasive Communication, *American Scientific Publishers*, 2006.

[9]. L. B. Ruiz, I. G. Siqueira, L. B Oliveria, H. C. Wong, J. M. S. Nogueira, A. A. Loureiro, Fault management in event-driven wireless sensor networks, in *Proceedings of the 7th International Workshop on Modeling Analysis and Simulation of Wireless and Mobile Systems*, 2004, pp. 149-156.

[10]. A. T. Tai, K. S. Tso, W. H. Sanders, Cluster-based failure detection service for large-scale ad hoc wireless network applications, in *Proceedings of the International Conference on Dependable Systems and Networks (DSN'04)*, 28 June-1 July 2004, pp. 805-814.

[11]. L. B. Ruiz, J. M. S. Nogueira, A. A. Loureiro, MANNA: A management architecture for wireless sensor networks, *Communications Magazine*, Vol. 41, Issue 2, 2003, pp. 116 -125.

[12]. M. Yu, H. Mokhtar, M. Merabti, Self-managed Fault Management in Wireless Sensor Networks, in *Proceedings of the 2nd International Conference on Mobile Ubiquitous Computing, Systems, Services and Technologies (UBICOMM '08)*, 2008. pp. 13-18.

[13]. W. L. Lee, A. Datta, R. Cardell-Oliver, WinMS: Wireless Sensor Network-management System, An Adaptive Policy-Based Management for Wireless Sensor Networks, *CSSE-06-001*, 2006.

[14]. S. Chessa, P. Santi, Crash faults identification in wireless sensor networks, *Computer Communications*, Vol. 25, No. 14, 2002, pp. 1273–1282.

[15]. B. Krishnamachari, S. Iyengar, Distributed Bayesian algorithms for fault-tolerant event region detection in wireless sensor networks, *IEEE Transactions on Computers*, Vol. 53, No. 3, March 2004, pp. 241-250.

[16]. W. Tsang-Yi, Y. S. Han, P. K. Varshney, P. Chen, Distributed fault-tolerant classification in wireless sensor networks, *IEEE Journal on Selected Areas in Communications*, Vol. 23, No. 4, 2005, pp. 724-734.

[17]. G. Venkataraman, S. Emmanuel, S. Thambipillai, A cluster-based approach to fault detection and recovery in wireless sensor networks, in *Proceedings of the 4th International Symposium on Wireless Communication Systems*, 2007, pp. 35-39.

[18]. R. Niu, P. K. Varshney, Distributed Detection and Fusion in a Large Wireless Sensor Network of Random Size, *Eurasip Journal on Wireless Communications and Networking*, Vol. 2005, No. 4, 2005, pp. 462-472.

[19]. H. C. Hsieh, J. -S. Leu, W. -K. Shih, A fault-tolerant scheme for an autonomous local wireless sensor network, *Computer Standards and Interfaces*, Vol. 32, 2010, pp. 215–221.

[19]. Z. Ying, X. Debao, Mobile Agent-based Policy Management for Wireless Sensor Networks, in *Proceedings of the International Conference on Wireless Communications, Networking and Mobile Computing (WCNM'05)*, 2005, pp. 1207-1210.

[20]. H. Huangshui, Q. Guihe, Fault Management Frameworks in Wireless Sensor Networks, in *Proceedings of the 4th International Conference on Intelligent Computation Technology and Automation*, 2011, pp. 1093 – 1096.

[21]. S. Mitra, A. De Sarkar, S. Roy, A Review of Fault Management System in Wireless Sensor Network, in *Proceedings of the CUBE International Information Technology Conference*, September 3–5, India, 2012, pp. 144-148.

[22]. I. Saleh, M. Eltoweissy, A. Agbaria, H. El-Sayed, A Fault Tolerance Management Framework for Wireless Sensor Networks, *Journal of Communications,* Vol. 2, No. 4, June 2007, pp.38-48.

[23]. S. Gobriel, S. Khattab, D. Moss´e, J. Brustoloni, R. Melhem, Ride Sharing: Fault Tolerant Aggregation in Sensor Networks Using Corrective Actions, in *Proceedings of the 3ʳᵈ Annual IEEE Communications Society Conference on Sensor, Mesh and Ad Hoc Communications and Networks, (SECON'06)*, Vol. 2, 2006, pp. 595-604.

[24]. S. Mitra, A. De Sarkar, Energy aware fault tolerant framework in Wireless Sensor Networks, in *Proceedings of the Applications and Innovations in Mobile Computing (AIMoC)*, 2014, pp. 139-145.

[25]. D. M. Beder J. Ueyama, M. L. Chaim, A generic policy-free framework for fault-tolerant systems: Experiments on WSNs, in *Proceedings of the IEEE International Conference on Networked Embedded Systems for Enterprise Applications*, 2011, pp. 1-7.

[26]. A. Bari, A. Jaekel, S. Bandyopadhyay, Optimal placement of relay nodes in two-tiered, fault tolerant sensor networks, in *Proceedings of the 12ᵗʰ IEEE Symposium on Computers and Communications (ISCC'07)*, 2007, pp. 159-164.

[27]. A. Bari, A. Jaekel, J. Jiang, Y. Xu, Design of fault tolerant wireless sensor networks satisfying survivability and lifetime requirements, *Computer Communications*, Vol. 35, 2012, pp. 320–330.

[28]. A. Kashyap, S. Khullerb, M. Shaymana, Relay placement for fault tolerance in wireless networks in higher dimensions, *Computational Geometry*, Vol. 44, Issue 4, 2011, pp. 206–215.

[29]. X. Cheng, D. -Z. Du, L. Wang, B. G. Xu, Relay sensor placement in wireless sensor networks, *Journal of Wireless Networks*, Vol. 14, Issue 3, 2008, pp. 347-355.

[30]. K. Dasgupta, M. Kukreja, K. Kalpaki, Topology-aware placement and role assignment for energy-efficient information gathering in sensor networks, in *Proceedings of the IEEE International Symposium on Computer and Communication*, 2003, pp. 341–348.

[31]. E. Falck, P. Floren, P. Kaski, J. Kohonen, P. Orponen, Balanced data gathering in energy-constrained sensor networks, *Algorithmic Aspects of Wireless Sensor Networks, Lecture Notes in Computer Science*, Vol. 3121, 2004, pp. 59–70.

[32]. J. Pan, Y. T. Hou, L. Cai, Y. Shi, S. X. Shen, Topology control for wireless sensor networks, in *Proceedings of the 9ᵗʰ Annual International Conference on Mobile Computing and Networking*, 2003, pp. 286–299.

[33]. G. Gupta, M. Younis, Fault-tolerant clustering of wireless sensor networks, in *Proceedings of the IEEE Wireless Communications and Networking Conference,* 2003, pp. 1579–1584.

[34]. G. Gupta, M. Younis, Load-balanced clustering of wireless sensor networks, in *Proceedings of the 38th Annual IEEE International Conference on Communications,* Vol. 3, 2003, pp. 1848–1852.

[35]. G. Gupta, M. Younis, Performance evaluation of load-balanced clustering of wireless sensor networks, in *Proceedings of the 10th International Conference on Telecommunications,* Vol. 2, 2003, pp. 1577–1583.

[36]. Y. T. Hou, Y. Shi, J. Pan, S. F. Midkiff, Lifetime-optimal data routing in wireless sensor networks without flow splitting, *Workshop on Broadband Advanced Sensor Networks*, San Jose, CA, 2004.

[37]. Y. T. Hou, Y. Shi, H. Sherali, S. F. Midkiff, On energy provisioning and relay node placement for wireless sensor networks, *IEEE Transactions on Wireless Communications*, Vol. 4, Issue 5, 2005, pp. 2579–2590.

[38]. J. Tang, B. Hao, A. Sen, Relay node placement in large scale wireless sensor networks, *Computer Communications*, Vol. 29, Issue 4, 2006, pp. 490–501.

[39]. R. Sokullu, L. O. Karaca, Comparative Study of Cross Layer Frameworks for Wireless Sensor Networks, in *Proceedings of the 1st International Conference on Wireless Communication, Vehicular Technology, Information Theory and Aerospace & Electronic Systems Technology, (Wireless VITAE'09),* May 17-20, Aalborg, Denmark, 2009, pp. 896 - 900.

[40]. R. Sokullu, L. O. Karaca, Simple and Efficient Cross-Layer Framework Concept for Wireless Sensor Networks, in *Proceedings of the 12th International Symposium on Wireless Personal Multimedia Communications,* Sept. 7 – 10, 2009, Sendai, Japan.

[41]. L. O. Karaca, Designing A Fault Tolerant Cross- Layer Framework in Wireless Sensor Networks, Ph. D Thesis, *Ege University*, Izmir, Turkey, 2010 (in Turkish).

Chapter 4

Hybrid RSS/AoA-based Localization of Target Nodes in a 3-D Wireless Sensor Network

Slavisa Tomic, Milica Marikj, Marko Beko, Rui Dinis, Milan Tuba

4.1. Introduction

Location aware services are becoming an integral part of many wireless systems. In a wireless sensor network (WSN), sensors are commonly deployed over a region of interest with limited to non-existing control of their location in space, *e.g.* thrown out of an airplane for sensing in hostile environments [1]. Installing a global positioning system (GPS) receiver in each device to determine its location is too expensive and limits the network applicability [2]. In order to sustain low implementation costs, only a small number of sensors are equipped with GPS devices (called anchors), while the remaining ones (called targets) determine their locations by using a kind of localization scheme that takes advantage of the known anchor locations [3].

Wireless localization schemes typically rely on range (distance) measurements. Depending on the available hardware, range measurements can be extracted from different characteristics of the radio signal, such as time-of-arrival (ToA), time-difference-of-arrival (TDoA), angle-of-arrival (AoA) or received signal strength (RSS) [4, 5]. Recently, hybrid systems that fuse two measurements of the radio signal have been investigated [6-13]. Hybrid systems profit by exploiting the benefits of combined measurements, *i.e.*, more available information. On the other hand, the price to pay for using such systems is the increased complexity of network devices which increases the network implementation costs [2, 4].

Approaches in [6-8] are based on the fusion of RSS and ToA measurements. A hybrid system that merges range and angle measurements was investigated in [9]. The authors in [9] proposed two estimators to solve the target localization problem in a 3-D scenario:

linear least squares (LS) and optimization based. The LS estimator is a relatively simple and well known estimator, while the optimization based estimator was solved by Davidson-Fletcher-Powell algorithm [14]. Although both estimators are simple to implement, they do not tightly approximate the maximum likelihood (ML) estimator, and their estimation accuracy can be further improved. In [10], the authors derived an LS and an ML estimator for the hybrid RSS/AoA localization problem. Their preliminary results showed that measurement fusion can significantly improve the estimation accuracy. A weighted LS (WLS) estimator for RSS/AoA localization problem was proposed in [12]. The authors determined the target location by using weighted ranges from two *nearest* anchors and a serving base station, together with an angular beam from the BS. In [10-12], authors investigated the hybrid RSS/AoA localization problem for 2-D scenario. A WLS estimator for 3-D RSS/AoA localization problem when the emitter's power is unknown was presented in [13]. However, the authors in [13] investigated a small-scale WSN, with low noise power.

In this work, we investigate the target localization problem in a 3-D WSN. A hybrid system that fuses distance and angle measurements, extracted from RSS and AoA information respectively, is employed. By using the RSS propagation model and simple geometry, we derive a novel objective function based on the WLS criterion. We show that the derived non-convex objective function can be transformed into a convex one, by using semidefinite programming (SDP) relaxation technique. Moreover, we show that the alteration of the proposed estimator to the case where the sensor's transmit power, P_T, is unknown is straightforward. Throughout our simulations, it was observed that the proposed SDP estimator, besides the location estimate, provides excellent estimation of P_T. Therefore, we take advantage of the power estimate and we propose a simple three-step procedure based on SDP to solve the target localization problem for the case of unknown P_T. Our simulation results confirm the effectiveness of the proposed estimators, which significantly reduce the estimation error in comparison with the state-of-the-art.

Throughout the work, upper-case bold type, lower-case bold type and regular type is used for matrices, vectors and scalars, respectively. $\mathbb{R}^n$ and $\mathbb{C}^n$ respectively denote the n dimensional real and complex Euclidean space. The operators $(\cdot)^T$ and $(\cdot)^H$ denote transpose and Hermitian, respectively. The normal (Gaussian) distribution with mean μ and variance σ^2 is denoted by $\mathcal{N}(\mu; \sigma^2)$. The N-dimensional identity matrix is denoted by I_N and the $M \times N$ matrix of all zeros by $0_{M \times N}$ (if no ambiguity can occur, subscripts are omitted). $\|x\|$ denotes the vector norm defined by $\|x\| = \sqrt{x^H x}$, where $x \in \mathbb{C}^n$. For Hermitian matrices A and B, $A \succcurlyeq B$ means that $A - B$ is positive semidefinite.

The remainder of this work is organized as follows. In Section 4.2, the RSS and AoA measurement models are introduced and the target localization problem is formulated. Section 4.3 presents the development of the proposed SDP estimators for both known and unknown P_T. In Sections 4.4 and 4.5, complexity and performance analysis are presented respectively, together with the relevant results in order to compare the performance of the newly presented estimators to the state-of-the-art. Finally, Section 4.6 summarizes the main conclusions.

4.2. Problem Formulation

We consider a WSN with N anchor nodes and one target node, where the locations of the anchor nodes are respectively denoted by $a_1, a_2, \ldots, a_N$, and the location of the unknown target node is denoted by x ($x, a_i \in \mathbb{R}^q, q = 3, i = 1, \ldots, N$). The location of the unknown target node is determined by using a hybrid system that fuses range and angle measurements. Combining two measurements of the radio signal is likely to improve the estimation accuracy, since it provides more information to the user.

Throughout this work, it is presumed that the range measurements are obtained from the RSS information exclusively. However, the RSS measurement model can be replaced with the path loss model by using the relationship $L_i = 10 \log_{10} \frac{P_T}{P_i}$ (dB), where L_i is the path loss between the unknown target and i-th anchor, P_T is the transmission power of a sensor node, and P_i is the received power at a distance $\|x - a_i\|$ from the transmitting sensor node [15, 16] as:

$$L_i = L_0 + 10\gamma \log_{10} \frac{\|x - a_i\|}{d_0} + n_i, \text{for } i = 1, \ldots, N, \qquad (4.1)$$

where L_0 denotes the path loss value at a short reference distance d_0 ($\|x - a_i\| \geq d_0$), γ is the path loss exponent (PLE), and n_i is the log-normal shadowing term modeled as a zero-mean Gaussian random variable with variance $\sigma_{n_i}^2$, i.e., $n_i \sim \mathcal{N}(0, \sigma_{n_i}^2)$. Furthermore, to obtain the AoA measurements (both azimuth and elevation angles) we assume that

73

either multiple antennas or a directional antenna is implemented at anchor nodes [9, 17].

As shown in Fig. 4.1, $\mathbf{x} = [x_1, x_2, x_3]^T$ and $\mathbf{a}_i = [a_{i1}, a_{i2}, a_{i3}]^T$ are respectively the unknown coordinates of the target and the known coordinates of the i-th anchor node, while d_i, ϕ_i and α_i represent the distance, azimuth angle and elevation angle between the target and the i-th anchor, respectively. From Fig. 4.1, we can see that it is possible that only one anchor node is sufficient to solve the localization problem in a 3-D environment when both range and angle measurements are available.

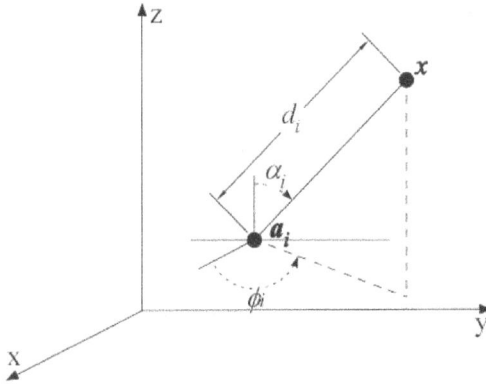

Fig. 4.1. Illustration of target and anchor locations in a 3-D space.

The ML estimate of the distance can be obtained from the RSS measurement model (4.1) as follows [4]:

$$\hat{d}_i = d_0 10^{\frac{L_i - L_0}{10\gamma}}, \text{ for } i = 1, \dots, N, \tag{4.2}$$

Applying simple geometry, azimuth and elevation angles measurements at anchor node $i = 1, \dots, N$ can be modeled respectively as [9]:

$$\phi_i = \arctan\left(\frac{x_2 - a_{i2}}{x_1 - a_{i1}}\right) + m_i, \tag{4.3}$$

and

$$\alpha_i = \arccos\left(\frac{x_3 - a_{i3}}{\|x - a_i\|}\right) + v_i, \tag{4.4}$$

where m_i and v_i are the measurement errors of azimuth and elevation angles respectively, modeled as zero mean Gaussian random variables, i.e., $m_i \sim \mathcal{N}(0, \sigma_{m_i}^2)$ and $v_i \sim \mathcal{N}(0, \sigma_{v_i}^2)$.

Given the observation vector $\boldsymbol{\theta} = [L_i, \phi_i, \alpha_i]^T$ $(\boldsymbol{\theta} \in \mathbb{R}^{3N})$, the conditional probability density function (PDF) is given as:

$$p(\boldsymbol{\theta}|\boldsymbol{x}) = \prod_{i=1}^{3N} \frac{1}{\sqrt{2\pi\sigma_i^2}} \exp\left\{-\frac{(\theta_i - f_i(\boldsymbol{x}))^2}{2\sigma_i^2}\right\}, \tag{4.5}$$

where

$$f(\boldsymbol{x}) = \begin{bmatrix} L_0 + 10\gamma \log_{10}\frac{\|x-a_1\|}{d_0} \\ \vdots \\ L_0 + 10\gamma \log_{10}\frac{\|x-a_N\|}{d_0} \\ \arctan\left(\frac{x_2-a_{12}}{x_1-a_{11}}\right) \\ \vdots \\ \arctan\left(\frac{x_2-a_{N2}}{x_1-a_{N1}}\right) \\ \arccos\left(\frac{x_3-a_{13}}{\|x-a_1\|}\right) \\ \vdots \\ \arccos\left(\frac{x_3-a_{N3}}{\|x-a_N\|}\right) \end{bmatrix}, \; \boldsymbol{\sigma} = \begin{bmatrix} \sigma_{n_1} \\ \vdots \\ \sigma_{n_N} \\ \sigma_{m_1} \\ \vdots \\ \sigma_{m_N} \\ \sigma_{v_1} \\ \vdots \\ \sigma_{v_N} \end{bmatrix}.$$

The most common estimator used in practice is the ML estimator, since it has the property of being asymptotically efficient (for large enough data records) [18]. The ML estimator forms its estimate as the vector $\hat{\boldsymbol{x}}$, which maximizes the conditional PDF in (4.5); hence, the ML estimator is obtained as:

$$\hat{\boldsymbol{x}} = \arg\min_{\boldsymbol{x}} \sum_{i=1}^{3N} \frac{1}{\sigma_i^2}[\theta_i - f_i(\boldsymbol{x})]^2. \tag{4.6}$$

Although the ML estimator is approximately the minimum variance unbiased estimator [18], the least squares (LS) problem in (4.6) is non-convex and has no closed-form solution. In the remainder of this work, we will show that the LS problem in (4.6) can be solved efficiently by applying a convex relaxation technique leading to an SDP estimator which can be solved efficiently by interior-point algorithms [19], for both cases of known and unknown P_T.

4.3. The Proposed SDP Estimators

In this section, we develop two estimators for both known and unknown P_T by using appropriate semidefinite cone relaxation techniques for 3-D target localization. For the sake of simplicity and without loss of generality, we assume that $\sigma_i = \sigma$ for $i = 1, ..., 3N$ in the rest of the work.

For sufficiently small noise, from (4.1) we have:

$$\lambda_i \|x - a_i\| \approx d_0, \tag{4.7}$$

where $\lambda_i = 10^{\frac{L_0 - L_i}{10\gamma}}$. Similarly, from (3) and (4), we get respectively:

$$c_i^T (x - a_i) \approx 0, \tag{4.8}$$

and

$$k_i^T (x - a_i) \approx \|x - a_i\| \cos(\alpha_i), \tag{4.9}$$

where $c_i = [-\tan(\phi_i), 1, 0]^T$, and $k_i = [0, 0, 1]^T$.

4.3.1. Transmit Power is Known

According to the WLS criterion and (4.7), (4.8) and (4.9), the target location estimate, $\hat{x}$, is found by minimizing the following objective function:

$$\hat{x} = \arg\min_x \sum_{i=1}^N w_i (\lambda_i \|x - a_i\| - d_0)^2 + \sum_{i=1}^N w_i \left(c_i^T (x - a_i) \right)^2 +$$
$$\sum_{i=1}^N w_i \left(k_i^T (x - a_i) - \|x - a_i\| \cos(\alpha_i) \right)^2, \tag{4.10}$$

where w_i is the weight defined as $w_i = 1 - \frac{\hat{d}_i}{\sum_{i=1}^N \hat{d}_i}$ for $i = 1, ..., N$. The motivation behind employing weights defined in the described manner is to give more importance to the measurements conducted from nearby anchors[1].

Problem in (4.10) is obviously non-convex and has no closed-form solution. To convert (4.10) into a convex problem, first, introduce auxiliary variables $r_i = \|x - a_i\|$ and $z = [z_i, g_i, h_i]^T$ ($z \in \mathbb{R}^{3N}$),

---

[1] Note that we can also derive the LS version of (10), *i.e.*, we can discard the weights. In our simulations, we have confirmed that this approach also provides a good solution.

where $z_i = w_i(\lambda_i \|x - a_i\| - d0)$, $\quad g_i = w_i(c_i^T(x - a_i))$, and
$h_i = w_i(k_i^T(x - a_i) - \|x - a_i\| \cos(\alpha_i))$, for $i = 1, \dots, N$. We get:

$$\min_{x,r,z} \|z\|^2$$

subject to

$$r_i = \|x - a_i\|, \text{ for } i = 1, \dots, N,$$

$$z_i = w_i(\lambda_i \|x - a_i\| - d0), \text{ for } i = 1, \dots, N,$$

$$g_i = w_i(c_i^T(x - a_i)), \text{ for } i = 1, \dots, N,$$

$$h_i = w_i(k_i^T(x - a_i) - \|x - a_i\| \cos(\alpha_i)), \text{ for } i = 1, \dots, N. \quad (4.11)$$

Introduce an epigraph variable t, and apply semidefinite cone constraint relaxation to obtain:

$$\min_{x,r,z,t} t$$

subject to

$$r_i \leq \|x - a_i\|, \text{ for } i = 1, \dots, N,$$

$$z_i = w_i(\lambda_i \|x - a_i\| - d0), \text{ for } i = 1, \dots, N,$$

$$g_i = w_i(c_i^T(x - a_i)), \text{ for } i = 1, \dots, N,$$

$$h_i = w_i(k_i^T(x - a_i) - \|x - a_i\| \cos(\alpha_i)), \text{ for } i = 1, \dots, N$$

$$\begin{bmatrix} I_{3N} & z \\ z & t \end{bmatrix} \succcurlyeq 0_{3N+1 \times 3N+1}. \quad (4.12)$$

The problem in (4.12) is a SDP problem, which can be efficiently solved by the CVX package [20] for specifying and solving convex programs. Note that we rewrote the constraint $\|z\|^2 \leq t$ into a semidefinite cone constraint form. In the further text, we will refer to (4.12) as "SDP1".

4.3.2. Transmit Power is Unknown

In order to minimize the expenses and achieve a low-cost localization system, testing and calibration of the devices are usually not the priority in practice [4]. This means that the node transmit power, P_T, is not calibrated, i.e., unknown. Unknowing P_T corresponds to unknowing L_0 in the path loss model; hence, in this subsection, L_0 is considered to be an unknown parameter that also needs to be estimated. Adaptation of the

SDP1 approach for known L_0 in (4.12) is straightforward for the case where L_0 is unknown. Notice that from (4.7) we can write:

$$\rho_i \|x - a_i\| \approx \eta d_0, \tag{4.13}$$

where $\rho_i = 10^{-\frac{L_i}{10\gamma}}$, and $\eta = 10^{-\frac{L_0}{10\gamma}}$ is an unknown parameter.

Then, by following the WLS principle and (4.13), (4.8) and (4.9), the below WLS problem is obtained:

$$\hat{x} = \arg\min_{x,\eta} \sum_{i=1}^{N} w_i (\rho_i\|x - a_i\| - \eta d_0)^2 + \sum_{i=1}^{N} w_i \left(c_i^T(x - a_i)\right)^2 +$$
$$\sum_{i=1}^{N} w_i \left(k_i^T(x - a_i) - \|x - a_i\|\cos(\alpha_i)\right)^2, \tag{4.14}$$

where weights are defined as $w_i = 1 - \frac{L_i}{\sum_{i=1}^{N} L_i}$ for $i = 1, \ldots, N$.

Applying similar procedure as in Section 4.3.1, the following SDP estimator is derived:

$$\min_{x,\eta,r,z,t} t$$

subject to

$$r_i \leq \|x - a_i\|, \text{ for } i = 1, \ldots, N,$$
$$z_i = w_i(\rho_i\|x - a_i\| - \eta d0), \text{ for } i = 1, \ldots, N,$$
$$g_i = w_i\left(c_i^T(x - a_i)\right), \text{ for } i = 1, \ldots, N,$$
$$h_i = w_i\left(k_i^T(x - a_i) - \|x - a_i\|\cos(\alpha_i)\right), \text{ for } i = 1, \ldots, N$$
$$\begin{bmatrix} I_{3N} & z \\ z & t \end{bmatrix} \succeq 0_{3N+1\times 3N+1}. \tag{4.15}$$

Even though the approach in (4.15) efficiently solves (4.6) for unknown L_0, we can further improve its performance. To do so, we will exploit the estimate of L_0, $\hat{L}_0$, which we obtain by solving (4.15), and we will take advantage of this estimate to solve another SDP problem as if L_0 is known. Hence, the proposed procedure for solving (4.6) when L_0 is unknown is summarized below:

1. Solve (4.15) to obtain the initial estimate of x, $\hat{x}'$;
2. Use $\hat{x}'$ to compute the ML estimate of L_0, $\hat{L}_0$ as:

$$\hat{L}_0 = \frac{\sum_{i=1}^{N}\left(L_i - 10\gamma \log_{10}\frac{\|\hat{\boldsymbol{x}}' - \boldsymbol{a}_i\|}{d_0}\right)}{N};$$

3. Exploit $\hat{L}_0$ to calculate $\hat{\lambda}_i = 10^{\frac{\hat{L}_0 - L_i}{10\gamma}}$, and use this estimate to solve the SDP in (4.12).

The main reason for applying this simple procedure is that, we observed in our simulations that after solving (4.15) we obtain an excellent ML estimation of L_0, $\hat{L}_0$, which is very close to the true value of L_0. This motivated us to take advantage of this estimated value and solve another SDP problem (4.12), as if L_0 is known. We denote this three-step procedure as "SDP2" in the further text.

4.4. Complexity Analysis

Table 4.1 provides an overview of the considered algorithms[2], together with their worst case computational complexities. Note that we investigated the worst case asymptotic complexity of the algorithms, *i.e.*, we present only the dominating elements in Table 4.1. The worst case computational complexity of the proposed SDP approaches was calculated according to [21], as:

$$O\left(\sqrt{L}\left(m\sum_{i=1}^{N_i}n_i^3 + m^2\sum_{i=1}^{N_i}n_i^2 + m^3\right)\right),$$

where L is the dimension of the SDP cone, given as a result of accumulating all SDP cones, m is the number of equality constraints, n_i is the dimension of the i-th semidefinite cone (SDC), and N_i is the number of SDC constraints [21].

From Table 4.1, one can see that the proposed SDP approaches are the most expensive in terms of the computational costs, as anticipated. However, the higher computational costs of the proposed approaches is

---

[2] Note that the WLS approach in [13] is not considered here. The reason is that after the WLS method was implemented for the scenarios considered in this work, where the area accommodating sensors and the noise power are much larger than in [13], it did not provide satisfactory results.

well justified by their superior performance in terms of the estimation accuracy, as we will see in the following text.

Table 4.1. Summary of the Considered Algorithms.

Algorithm	Description	Complexity
SDP1	The proposed SDP estimator for known PT in (4.12)	$O(N^{4.5})$
SDP2	The proposed SDP approach for unknown PT in Section 3.2	$2 \cdot O(N^{4.5})$
LS	The LS estimator for known PT proposed in [9]	$O(N)$

4.5. Simulation Results

In this section, computer simulations are performed in order to compare the performance of the proposed approach with the state-of-the-art. The proposed algorithm was solved by using the MATLAB package CVX [20], where the solver is SeDuMi [22].

To generate the measurements, equations (4.1), (4.3) and (4.4) were used. We considered a random deployment of nodes inside a square region of length $B = 30$ m in each Monte Carlo (M_c) run. Random deployment of nodes is of practical interest, since the algorithms are tested against various network topologies. The PLE was set to $\gamma = 2$, the reference distance $d_0 = 1$ m, and the reference path loss $L_0 = 40$ dB. As the main performance metric we used the root mean square error (RMSE), defined as

$$\text{RMSE} = \sqrt{\sum_{i=1}^{M_c} \frac{\|x_i - \hat{x}_i\|}{M_c}},$$

where $\hat{x}_i$ denotes the estimate of the true target location, x_i, in the i-th M_c run.

In order to demonstrate the superiority of a hybrid RSS/AoA system over the RSS one, we call the reader's attention to Fig. 4.2 below. Fig. 4.2 illustrates the cumulative distribution function (CDF) comparison of the mismatch error (ME) in the target location estimation of the proposed SDP1 approach and the SDP1 approach when RSS measurements are employed exclusively, SDP1_{RSS}, for $N = 6$ and $\sigma = 6$ (dB, deg), where the last notation stands for $\sigma = 6$ dB for RSS measurements and $\sigma = 6$ degrees for AoA measurements. We define the ME as a vector

$\mathbf{ME} = [\mathrm{ME}_i]$, where each element of the vector is $\mathrm{ME}_i = \|x_i - \hat{x}_i\|$ (m), for $i = 1, ..., M_c$. From Fig. 4.2, it is obvious that fusing two measurements of the radio signal, in this case RSS and AoA, significantly enhances the estimation accuracy of the proposed SDP approach in comparison with employing only one (RSS) measurement. To be more specific, we can see that the proposed hybrid approach achieves $\mathbf{ME} \leq 5$ m in more than 90 % of the cases, while the SDP1$_{\mathrm{RSS}}$ achieves the same performance in not more than 10 % of the cases. This behavior is not surprising, since more information is obtained by combining two radio measurements. However, acquiring two measurements usually comes with an additional cost resulting from the increased complexity of the network devices.

Fig. 4.3 illustrates the RMSE versus σ comparison of the considered algorithms, when $N = 4$. From Fig. 4.3, we can see that the performance of all considered approaches deteriorates as σ grows, as expected. Moreover, it is seen that the performance gap between approaches increases with σ. The proposed approach for known P_T outperforms the state-of-the-art for all choices of σ, reducing the estimation error for roughly 7 m for $\sigma = 6$ (dB, deg). Moreover, the proposed approach for unknown P_T performs better than the LS method, reducing the estimation error for about 3.5 m for $\sigma = 6$ (dB, deg).

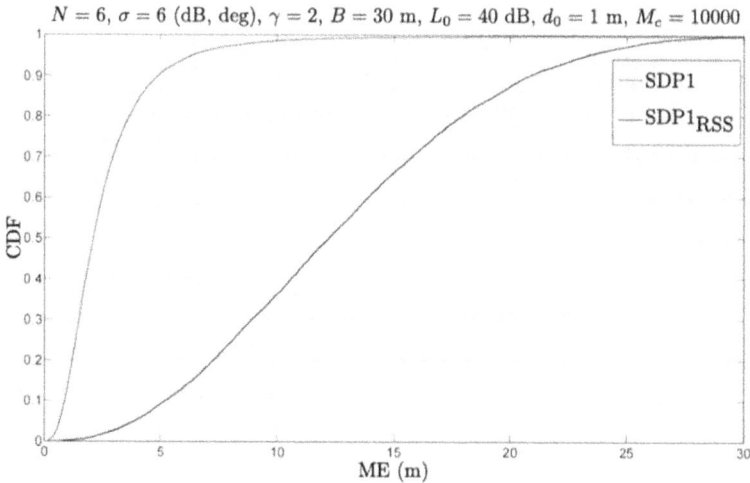

Fig. 4.2. CDF versus ME (m) comparison, when $N = 6, \sigma = 6$ (Db, deg), $\gamma = 2, B = 30$ m, $L_0 = 40$ dB, $d_0 = 1$ m, $M_c = 10000$.

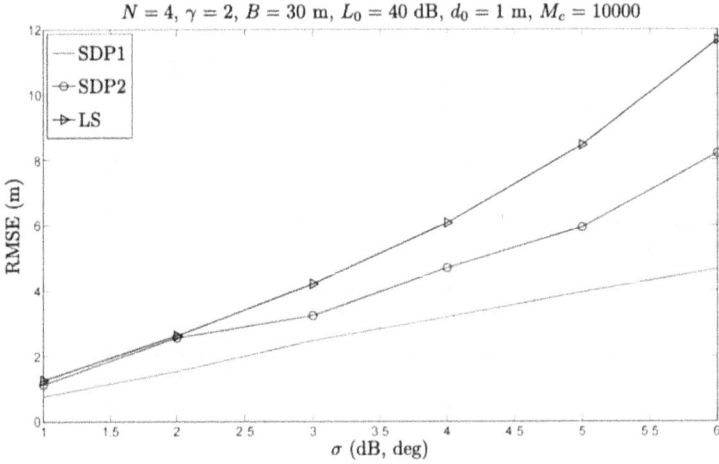

Fig. 4.3. RMSE versus σ (dB, deg) comparison, when $N = 4$, $\gamma = 2, B = 30$ m, $L_0 = 40$ dB, $d_0 = 1$ m, $M_c = 10000$.

Fig. 4.4 illustrates the RMSE versus σ comparison of the considered algorithms, when $N = 6$. From Fig. 4.4, one can see that the proposed approaches outperform the existing one for all choices of σ, improving the estimation accuracy for about 6 m (SDP1) and 4 m (SDP2), when $\sigma = 6$ (dB, deg).

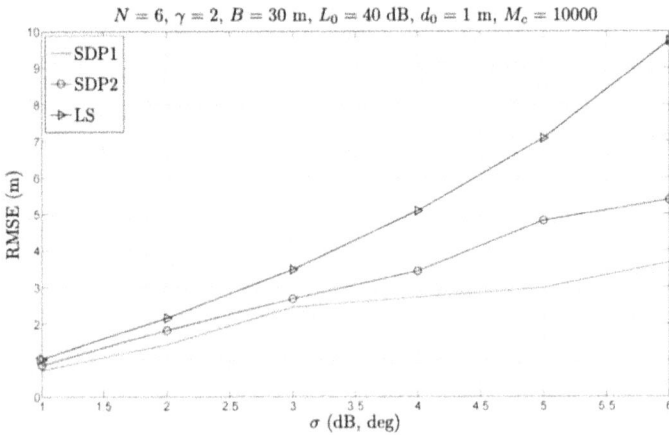

Fig. 4.4. RMSE versus σ (dB, deg) comparison, when $N = 6$, $\gamma = 2, B = 30$ m, $L_0 = 40$ dB, $d_0 = 1$ m, $M_c = 10000$.

Fig. 4.5 illustrates the RMSE versus N comparison of the considered algorithms, when $\sigma = 6$ (dB, deg). As predicted, the estimation error decreases as N increases, as we can see from Fig. 4.5. From the figure, it is clear that the new approaches outperform considerably the existing one for all choices of N. The biggest performance margin is noticeable for low N, and the margin slowly reduces as N grows. This behavior is not surprising, since with the increase of N the information gathered inside the network becomes sufficient to allow good performance for all estimators. Finally, even though we derived our estimators under the assumption that the noise is small, Fig. 4.5 exhibits excellent performance of the proposed approaches in high-noise environments.

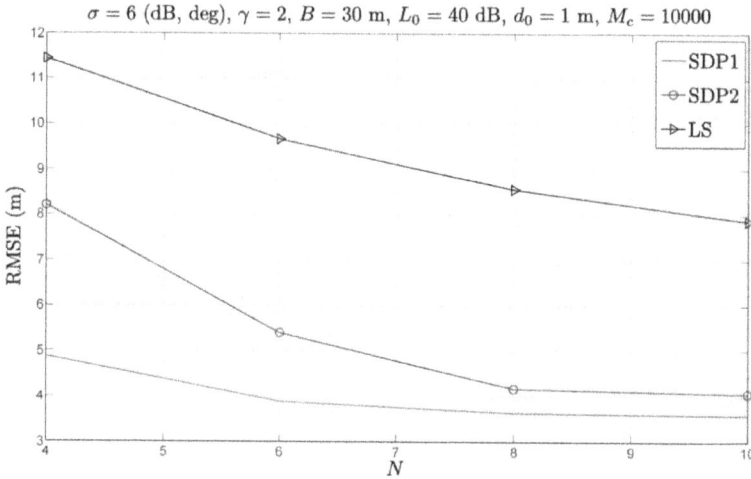

Fig. 4.5. RMSE versus N comparison, when $\sigma = 6$ (dB, deg), $\gamma = 2, B = 30$ m, $L_0 = 40$ dB, $d_0 = 1$ m, $M_c = 10000$.

4.6. Conclusions

In this work, we investigated a hybrid localization system which combines range and angle measurements for target node localization in a 3-D WSN. We proposed a novel objective function based on the LS criterion, and we derived two estimators based on SDP relaxation technique that tightly approximate the ML estimator for small noise, for both known and unknown P_T. Although computationally more demanding than the state-of-the-art approach, the simulation results confirmed their superiority in terms of the estimation accuracy. We have investigated various scenarios of the localization problem, where all sensor nodes where deployed randomly. For any setting considered in this work, the new approaches outperform considerably the state-of-the-art.

Acknowledgements

This work was partially supported by Fundação para a Ciência e a Tecnologia under Projects, UID/EEA/50008/2013 (IT pluriannual founding and HETNET), EXPL/EEI-TEL/0969/2013 MANY2COMWIN, EXPL/EEI-TEL/1582/2013 GLANC, and UID/EEA/00066/2013 (CTS, UNINOVA pluriannual founding), EnAcoMIMOCo EXPL/EEITEL/2408/2013, ADIN PTDC/EEI-TEL/2990/2012, PTDC/EEITEL/6308/2014-HAMLeT, and the grant SFRH/BD/91126/2012 and Ciência 2008 Post-Doctoral Research grant.

References

[1]. L. Buttyán and J. P. Hubaux, Security and Cooperation in Wireless Networks: Thwarting Malicious and Selfish Behavior in the Age of Ubiquitous Computing, *Cambridge University Press*, New York, NY, USA, 2007.

[2]. N. Patwari, J. N. Ash, S. Kyperountas, A. O. Hero III, R. L. Moses, and N. S. Correal, Locating the Nodes: Cooperative Localization in Wireless Sensor Networks, *IEEE Sig. Process. Mag.*, Vol. 22, No. 4, July 2005, pp. 54-69.

[3]. G. Destino, Positioning in Wireless Networks: Noncooperative and Cooperative Algorithms, PhD Thesis, *University of Oulu*, Oulu, Finland, 2012.

[4]. N. Patwari, Location estimation in sensor networks, PhD Thesis, *University of Michigan*, Ann Arbor, MI, USA, 2005.

[5]. S. Tomic, M. Beko, and R. Dinis, RSS-based Localization in Wireless Sensor Networks Using Convex Relaxation: Noncooperative and Cooperative Schemes, *IEEE Trans. Vehic. Technol.*, 64, 5, 2015, pp. 2037-2050.

[6]. A. Bahillo, S. Mazuelas, R. M. Lorenzo, P. Fernández, J. Prieto, R. J. Durán, and E. J. Abril, Hybrid RSS-RTT Localization Scheme for Indoor Wireless Networks, *EURASIP J. Advan. Sig. Process.*, Vol. 2010, No. 1, Mar. 2010, pp. 1–12.

[7]. U. Hatthasin, S. Thainimit, K. Vibhatavanij, N. Premasathian, and D. Worasawate, The Use of RTOF and RSS for a One Base Station RFID system, *IJCSNS*, Vol. 10, No. 7, July 2010, pp. 862–865.

[8]. T. Gädeke, J. Schmid, M. Krüger, J. Jany, W. Stork, and K. D. Müller-Glaser, A Bi-Modal Ad-Hoc Localization Scheme for Wireless Networks Based on RSS and ToF Fusion, in *Proceedings of the WPNC*, pp. 1–6, Mar. 2013.

[9]. K. Yu, 3-D Localization Error Analysis in Wireless Networks, *IEEE Trans. Wirel. Commun.*, Vol. 6, No. 10, Oct. 2007, pp. 3473–3481.

[10]. S. Wang, B. R. Jackson, and R. Inkol, Hybrid RSS/AOA Emitter Location Estimation Based on Least Squares and Maximum Likelihood Criteria, in *Proceedings of the IEEE QBSC*, pp. 24–29, Jun. 2012.

[11]. L. Gazzah, L. Najjar, and H. Besbes, Selective Hybrid RSS/AOA Approximate Maximum Likelihood Mobile intra cell Localization, in *Proceedings of the European Wireless Conference*, Apr. 2013.

[12]. L. Gazzah, L. Najjar, and H. Besbes, Selective Hybrid RSS/AOA Weighting Algorithm for NLOS Intra Cell Localization, in *Proceedings of the IEEE WCNC*, Apr. 2014, pp. 2546–2551.

[13]. Y. T. Chan, F. Chan, W. Read, B. R. Jackson, and B. H. Lee, Hybrid Localization of an Emitter by Combining Angle-of-Arrival and Received Signal Strength Measurements, in *Proceedings of the IEEE CCECE*, May 2014, pp. 1–5.

[14]. R. Fletcher, Practical Methods of Optimization, *John Wiley & Sons*, Chichester, UK, 1987.

[15]. T. S. Rappaport, Wireless Communications: Principles and Practice. *Prentice-Hall*, Upper Saddle River, NJ, USA, 1996.

[16]. M. L. Sichitiu and V. Ramadurai, Localization of Wireless Sensor Networks with a Mobile Beacon, in *Proceedings of the IEEE MASS*, Oct. 2004, pp. 174–183.

[17]. Z. Xiang and U. Ozguner, A 3-D Positioning System for Off-road Autonomous Vehicles, in *Proceedings of the IEEE International Conference on Intelligent Vehicle Symposium*, Jun. 2005, pp. 130-135.

[18]. S. M. Kay, Fundamentals of Statistical Signal Processing: Estimation Theory, *Prentice-Hall*, Upper Saddle River, NJ, USA, 1993.

[19]. S. Boyd and L. Vandenberghe, Convex Optimization, *Cambridge University Press*, New York, NY, USA, 2004.

[20]. M. Grant and S. Boyd, CVX: Matlab Software for Disciplined Convex Programming. Version 1.21. Available online: http://cvxr.com/cvx (accessed on 15 April 2010).

[21]. I. Pólik and T. Terlaky, Interior Point Methods for Nonlinear Optimization, Nonlinear Optimization, G. Di Pillo, F. Schoen, Eds., 1st Edition, *Springer*, 2010.

[22]. J. F. Sturm, Using SeDuMi 1.02, a MATLAB Toolbox for Optimization Over Symmetric Cones, *Optimization Methods and Software*, Vol. 11, No. 1–4, 1999, pp. 625–653.

Chapter 5

Cost-effective Early Breast Tumor Detection Imaging System: Tactile and Near-Infrared Hybrid Imaging

Jong-Ha Lee

5.1. Introduction

Traditionally, physicians have used palpation to detect breast tumors or prostate tumors, which is based on the observation that the tissue abnormalities are usually associated with localized changes in mechanical properties such as low elasticity and stiffer tissues [1, 2]. To help physicians detect tumors more efficiently, various imaging techniques utilizing different imaging modalities such as computer tomography, ultrasonic imaging, nuclear magnetic resonance imaging, and x-rays have been developed [3-5]. However, each of these techniques has limitations, including the radiation to the body, low specificity, complicated system, etc. Moreover, these techniques can only provide the spatial information of the tumor. They do not measure mechanical properties directly. The absolute material properties are very important to measure the severity of the tumor. Identifying a stiff region relative to the surrounding region does not lead to diagnosing tissue abnormalities completely. Therefore it is desirable to measure the absolute elastic modulus directly using tactile elasticity imaging technique.

In fact, different tactile sensors using diverse approaches have already been investigated in robotic systems and medical tools for surgery. They are based on piezoelectric [6-8], piezoresistive [9, 10], or capacitive sensing [11]. Some sensors provide good spatial resolution through the use of microelectromechanical systems (MEMS) technology. However, its small measurable force range due to the brittle sensing elements such as silicone based diaphragms has not proven to be a reliable biomedical tool. In addition, most of them are in the form of an array of distributed pressure sensors on a flat plate and merely detect the applied force at that point. Without the ability to measure the displacement of the tissue deformation, the sensor cannot estimate the elasticity. Elasticity is used in cancer detection. Recently, some research groups use force sensing resistors and a super-resolution algorithm for a neck palpation device [12]. However, this approach can only detect the relative stiffness, not the absolute elastic modulus. Even though spatial resolution of the tactile sensor has been improved by the super-resolution algorithm, it has still fairly low resolution compared to human fingers, which were millions of mechanoreceptors per square inch of the skin. Some tactile sensors use piezoelectric cantilevers for the absolute elasticity measurement [13, 14]. However, this method requires auxiliary instruments such as oscilloscope or voltage generators. This scheme also has relatively low spatial resolution due to its large size of the probe. Therefore in order for tactile sensors to be successfully developed as the palpation tool, high tactile spatial resolution is necessary for the precise elasticity measurement.

In this chapter, we present a newly designed tactile elasticity imaging sensor. The rigid waveguide transduction based optical tactile sensors are already investigated in [15, 16]. Our system is inspired by this system with important differences. In the current design, a polydimethylsiloxane (PDMS) are used to make a multi-layer flexible transparent waveguide. The mechanical properties (i.e. elastic modulus) of each layer have been matched with the three human finger layers, dermis, epidermis, and subcutanea, to maximize the sensitivity of touch. In order to have high tactile spatial resolution, we utilize the total internal reflection principle in the waveguide. A force applied to a waveguide causes the light to change the critical angle of internally reflected lights, and results in light scattering which can be captured by a camera. The salient feature of this sensor compared to the other tactile sensors is the capability of measuring the elasticity of the contacted object without any external force sensor. In the current design, the force distribution has been measured through the integration of tactile image pixel values. In order to accurately estimate 3-D displacements of the contacted object deformation, a non-rigid pattern matching algorithm is developed. This technique relies on matching the random patterns recorded in tactile images to obtain the surface displacements and gradients from which the strain field can be determined. The obtained stress and strain information are finally used to identify the elasticity of the contacted object.

In the following section, the background of human tactile perception is given. Next, the design and characteristic of the tactile elasticity imaging

88

sensor is presented and its sensing principle is introduced. Then, the stress estimation and non-rigid pattern matching algorithm for the strain estimation are discussed. Finally, the experimental results and conclusions are presented.

5.2. Background

The tactile elasticity imaging sensor emulates a human finger. Here, we briefly review the human tactile perception and human finger biology.

5.2.1. Human Tactile Perception

Touch sensation is perceived via physical contact mainly through the skin. Human skin has about five million sensory cells, however, the cells are not evenly distributed. Areas such as fingertips and lips are more sensitive to touch because they have more nerve endings. Fingers can perceive a wide variety of tactile information such as roughness, softness, humidity, temperature, friction, pain, vibration and hardness. Human fingers also have the amazing ability to detect inclusion, such as tumors inside the tissues. In general, sensory receptors can be classified by their functions: chemoreceptors (chemical stimuli), nociceptors (pain), osmoreceptors (osmolarity of fluids), photoreceptors (light stimuli), and mechanoreceptors (mechanical stimuli) [17]. There are also two sensory systems that react via contact with a physical object: exteroceptive and proprioceptive sensory systems. Proprioceptive system is the sense of the relative position of neighboring parts of the body. Exteroceptive system is the response to external stimuli such as temperature, deformation of the skin and mechanical stimuli. The proposed sensor is mainly dealing with mechanoreceptors and exteroceptive sensor.

5.2.2. Biology of the Human Finger

A human finger has an oval shaped cross section, composed of tissue, and nail. The curved surface allows consistent and precise grasping and manipulation. Nails are effective in enlarging the stimuli on mechanoreceptors by sandwiching the tissue between the surface and the nail. Human tissues are made up of multiple layers: epidermis, dermis and subcutanea. Each layer has different physical properties. The outmost layer is the epidermis (elastic modulus: 1.4×10^2 Pa); beneath it is the dermis layer (8.0×10^4 Pa) and the layer closest to the bone is the subcutanea (3.4×10^4 Pa) [16]. The epidermis is the hardest layer, with the smallest elasticity at approximately 1 mm thickness. The dermis is a softer layer with more elasticity, usually 1 to 3 mm thick. The subcutanea, which fills the space between the dermis and bone, is mainly composed of fat and functions as a cushion when shock load is applied to the finger. Due to the differences in elastic coefficients, there is greater deformation of the inner layers, dermis and subcutanea, than the outmost layer, epidermis, when the finger presses into or moves along a surface. The multi-layer structure enhances the effective texture and hardness perception, which is why we have emulated this multi-layer structure for the proposed sensor.

5.3. Sensor Design and Sensing Principle

In this section, we present the concept, fabrication, and characterization of our sensor in detail.

5.3.1. Design Requirement for Emulating Human Finger

The tactile elasticity imaging sensor emulates the structure of a human finger. The design requirement is as follows.

1) *Human tissues*: Polydimethylsiloxane (PDMS) is used for emulating human tissue. PDMS creates a soft contact surface, which has proven to be effective in detecting the texture of material.

2) *Three-layered structure*: Emulating the structure of human tissue, three types of the PDMS with different elasticity are stacked together. This allows for more sensitive perception.

3) *Distribution of bone and nail elements*: In order to effectively obtain tactile sensory data, parts that function as the bone and nail are situated at the base of the sensor. In the current design, a heat-resistant borosilicate glass plate is used as the substrate for the stacked PDMS.

4) *Distributed sensor elements*: To emulate mechanoreceptors of a human finger, an optical based sensing method using a light reflection pattern and a digital imager is used. This is to obtain high spatial distribution of contact force.

5.3.2. Sensor Design

Fig. 5.1(a) shows the schematic of the tactile sensor module and Fig. 5.1(b) shows the integrated tactile sensor. The elasticity imaging sensor comprises of an optical waveguide, light sources, and a digital imager.

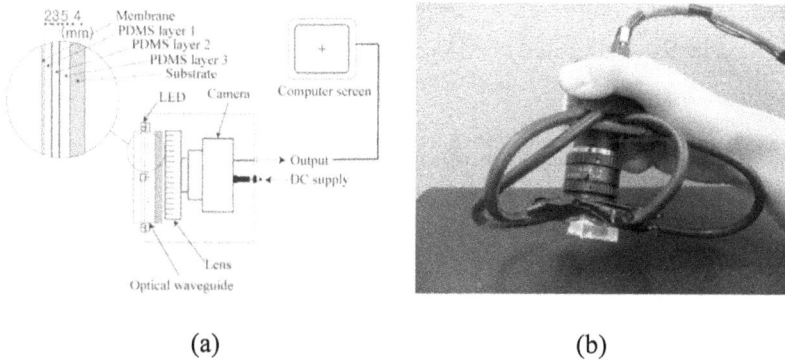

(a)   (b)

Fig. 5.1. (a) Schematic of the tactile elasticity imaging sensor;
(b) The integrated version.

The optical waveguide is the main sensing probe of the device. The optical waveguide is composed of PDMS (PDMS, $Si(CH_3)_2$), which is a high performance silicone elastomer [18, 19]. The optical waveguide needs to be transparent and PDMS meets this requirement. In the current design, one of the Hydroxyl-terminated PDMS, RTV6186 has been used (R.S. Hughes, Baltimore, MD). The PDMS is produced through a process of pouring viscous fluid silicone and a catalyst into a mold cavity. Here, the viscous fluid silicone used is vinyl-stopped phenylmethypolyer and the catalyst is a mixture of components, including methylhydrogen polysiloxane, dimethyl, methyvinyl siloxane and dimethylvinyl terminated. The viscous fluid silicone is hardened by the catalyst. The hardness is dependent on the ratio of silicone and catalyst. The elastic moduli of three PDMS layers are set as the modulus values of epidermis, dermis, and subcutanea. The height of each layer is 2 mm for epidermis layer (PDMS layer 1), 3 mm for dermis layer (PDMS layer 2) and 5 mm for subcutanea layer (PDMS layer 3), respectively. The fabricated PDMS optical waveguide is shown in Fig. 5.2.

91

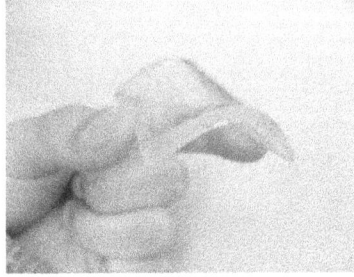

Fig. 5.2. Fabricated PDMS optical waveguide. The waveguide
is elastic and flexible.

The digital imager is a mono-cooled complementary camera with
4.65 μm (H) × 4.65 μm (V) individual pixel size (FLEA2, Point Grey
Research, British Columbia). The maximum lens resolution is 1392 (H)
× 1042 (V) with the angle of view is 60°. The camera is placed below
an optical waveguide. A heat-resistant borosilicate glass plate is placed
between the camera and the optical waveguide to sustain an optical
waveguide without losing camera resolution. The glass also functions as
the bone and nail in a human finger.

The internal light source is a micro-LED (Unique-Leds, Newalla, OK)
with a diameter of 1.8 mm. There are four LEDs on the four sides of the
waveguide to provide illumination. The direction and incident angle of
the LED light has been calibrated with the cone of acceptance angle and
is described in the next section.

5.3.3. Sensing Principle

Fig. 5.3 illustrates the conceptual diagram of the sensing principle. The
tactile elasticity imaging sensor is developed based on the optical
phenomenon known as total internal reflection (TIR) of light within an
optical waveguide.

If two mediums have different indices of refraction, and the light is shone
through those two mediums, then a fraction of light is transmitted and
the rest is reflected. The amount of reflection is dependent on the angle
of incidence. There is a critical angle above which the ray is completely
reflected. The basic principle of the sensor system lies in the monitoring
of the reflected light caused by the changing of the critical angle by
contact. The intensity of the reflected light is related to the applied force

and the strain on the optical waveguide. Here we investigate TIR in the multi-layer optical waveguide using ray optics approximation.

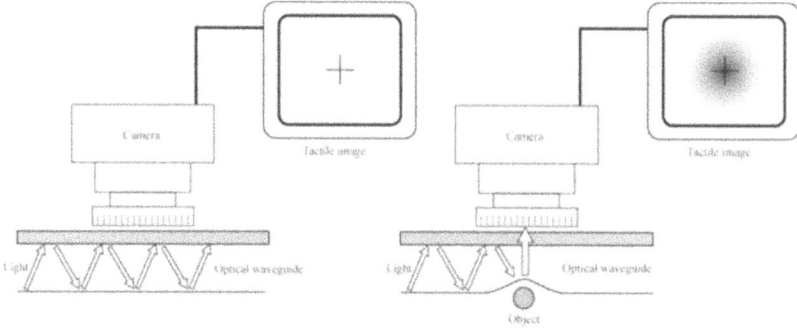

Fig. 5.3. Schematic diagram of the sensing principle. The light strays outside the waveguide as the optical waveguide deforms according to the applied force.

Consider light trapped inside the waveguide in the geometry shown in Fig. 5.4. The basic design of the optical waveguide plate consists of three different refractive indices of PDMS. Consider three PDMS layers that are non-absorbing mediums (refractive index: n_1, n_2, n_3) on a heat resistant borosilicate glass plate (refractive index: n_4). The magnitude of refractive index is set as the highest at medium 4 and decreases toward to medium 1, i.e., $n_1 < n_2 < n_3 < n_4$. Medium 0 and medium 5 are air which is the absorption-free medium, i.e., $n_0 = n_5 = 1$. Assume that LED light sources are placed around the middle of the PDMS layers. Light is incident from the outside of PDMS layers and strikes each layer of the PDMS stack. Due to Snell's law the propagation angles γ_i in each layer i, $i=1, 2, 3, 4$ are bound by the following relations.

$$n_1 \sin \gamma_1 = n_0 \sin \gamma_0$$

$$n_2 \sin \gamma_2 = n_1 \sin \gamma_1$$

$$n_3 \sin \gamma_3 = n_2 \sin \gamma_2 \qquad (5.1)$$

$$n_4 \sin \gamma_4 = n_3 \sin \gamma_3$$

$$n_5 \sin \gamma_5 = n_4 \sin \gamma_4$$

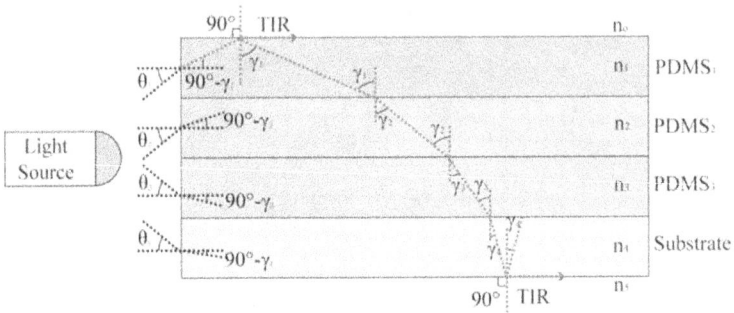

Fig. 5.4. Ray propagation inside the multi-layer optical waveguide under total internal reflection (TIR).

Here n_0 and n_5 are the refractive indices of air $n_0 = n_5 = 1$, and the critical TIR condition has been achieved when $\gamma_0 = \gamma_5 = 90^o$ at the boundaries with air. Light propagating in the waveguide with angles γ_1, γ_2, γ_3, γ_4 or higher in their respective layers will remain trapped inside the waveguide. The critical angle indicates the minimum propagation angle for TIR. To make the propagation angle below the critical angle in a waveguide, the acceptance angle for light sources has been calculated.

The acceptance angle is the maximum angle, under which light directed into the waveguide remains trapped inside. Angles γ_1, γ_2, γ_3, γ_4 are related to the acceptance angle θ_i in each layer i by the Snell's Law:

$$\sin \theta_i = n_i \sin(\gamma_0 - \gamma_i) = n_i \sin(\gamma_5 - \gamma_i) = n_i \sin(90^o - \gamma_i) = n_i \cos \gamma_i. \tag{5.2}$$

Further, transforming Eq. (5.2), we obtain

$$\sin \theta_i = n_i \cos \gamma_i = n_i (1 - \sin^2 \gamma_i)^{1/2} = (n_i^2 - n_i^2 \sin^2 \gamma_i)^{1/2}. \tag{5.3}$$

It follows from Eq. (5.1) that all $n_i \sin \gamma_i$ are equal to $n_0 \sin 90^o$, which is equal to 1 for air. Therefore, we finally have

$$\theta_i = \sin[(n_i^2 - 1)^{1/2}]^{-1} \tag{5.4}$$

for each layer i. Light, incident on layer i under angle θ_i will be trapped inside the waveguide. For instance, if n_1, n_2, n_3, n_4 are measured approximately 1.38, 1.39, 1.40, 1.41, the acceptance angles, θ_i, are calculated as 71.98^o, 74.89^o, 78.46^o, 83.73^o, respectively. Thus, for the TIR in the waveguide, the spatial radiation pattern of LED should be smaller than $71.98^o \times 2 = 143.96^o$. Fig. 5.5 shows the total internal reflection in a three layer PDMS optical waveguide using four LED light sources.

Fig. 5.5. The total internal reflection in a three layer PDMS optical waveguide using four LED light sources.

5.3.4. Sensor Specification

Spatial resolution between sensing points: The resolution of the tactile elasticity imaging sensor is going to be the pixel size of the camera. The spatial resolution between sensing points of the fingertip is at least 0.1 mm, which translates into an approximately 200×300 elements grid on a fingertip-sized area (20 mm $\times$ 30 mm) [18]. In the current design, the pattern discrimination ability of the elasticity-imaging sensor is 4.65 µm and translates into an approximately 4301×6415 elements grid on a fingertip-sized area.

Temporal resolution: With regard to the human fingertip temporal resolution, the vibration bandwidth reported at the fingertips is a few Hz for separate touches and hundred Hz for continuous sensing. The camera that we chose had a 1392×1042 resolution at 80 frames per second

(80 Hz). However, this temporal resolution can be improved depending on the camera.

Force sensitivity: Sensitivity is described in terms of the relationship between the physical signal (input) and the electrical signal (output) and is generally the ratio between a small change in the input to a small change in the output signal. The force sensitivity of the proposed tactile sensor force is approximately 2.5×10^{-3} N compared to the fingertip force sensitivity of 2.0×10^{-2} N [21].

Linearity/hysteresis: The human skin response has hysteresis. It has been noted that, for example, the force required maintaining a given indentation on the skin decreases as the probe is held against the skin [16]. The skin relaxes with time, with an observed length of up to 8 seconds. The proposed sensor is stable, repeatable and continuous in its variable output signal. The response of the sensor is non-hysteric and verified in the experiments.

5.4. Elasticity Estimation

The elasticity is obtained using stress and strain information from tactile images. Since the stress is measured as force per unit area, we estimate the applied force by contacting the object using the integrated pixel value of tactile image. For the strain information, the displacement of a material deformation under the applied force has been measured by tracking the control points extracted from two different tactile images. In this section, stress and strain estimation algorithms are been discussed in detail.

5.4.1. Stress Estimation

In this section, we present the stress estimation method using the obtained tactile image. If the optical waveguide of a sensor is compressed by the contacting object, it is deformed in both compressive and shearing directions. Because the light scatters at the contact area, the pixel value of the tactile image acquired by the camera distributes as a Gaussian function, in which the pixel intensity is the highest at the centroid and decreases with increased distance from the centroid [15].

Table 5.1. Sensory specification of the human fingertip and tactile elasticity imaging sensor.

Design criteria	Human fingertip	Tactile elasticity imaging sensor
Spatial resolution between sensing points	0.1 mm	4.65 μ m
Temporal resolution	0~100 Hz	0~80 Hz
Force sensitivity	2.5×10^{-3} N	2.0×10^{-2} N
Hysteresis	High	Low

Let $I(x,y)$ be the pixel value of the image plane. Since $I(x,y)$ is proportional to the contact stress, $P(x,y)$, caused by the contact between optical waveguide and the object, it can be expressed as follows:

$$P(x,y) = f(I(x,y)), \tag{5.5}$$

where f is the conversion function which is determined by experiments. If S is designated as the contact area between the optical waveguide and the contact object, then the vertical force F_z is obtained by integrating the stress over the contact area as follows

$$F_z = \int_S P(x,y)dS. \tag{5.6}$$

In order to determine horizontal force vectors F_x and F_y, the x - and y - coordinates of the centroid (X_c, Y_c), are calculated by

$$X_c = \int_S I(x,y)x dS / \int_S I(x,y)dS, \tag{5.7}$$

$$Y_c = \int_S I(x,y)y dS / \int_S I(x,y)dS. \tag{5.8}$$

The movements of the x - and y - components of the centroid are denoted as u_x and u_y and expressed as

$$u_x = X_c^{(t)} - X_c^{(t-1)}, \tag{5.9}$$

$$u_y = Y_c^{(t)} - Y_c^{(t-1)}. \tag{5.10}$$

where t and $t-1$ represent current and prior steps, respectively. If friction between the optical waveguide and contact object is ignored, then x - and y - directional forces F_x and F_y are calculated as follows

$$F_x = K_x u_x,$$ (5.11)

$$F_y = K_y u_y.$$ (5.12)

where K_x and K_y are x - and y - directional spring constants of the optical waveguide, respectively. Spring constants are determined experimentally. If we calculate the applied force, then the stress is calculated from the applied force per unit contact area.

5.4.2. Strain Estimation using Non-rigid Pattern Matching Algorithm

The strain measurement of a material under loading is achieved by tracking the displacement of control points extracted from series of tactile images. This concept is attractive, but manual measurement of control point positions is tedious and subject to error. This limits the number of control points that can be used and measured, and the spatial resolution of displacement fields. In this chapter, to tracking the control points efficiently and automatically, non-rigid pattern matching algorithm is developed. The essence of this algorithm is to automatically measure displacements by tracking the change in position of control points. Fig. 5.6 represents the concept of tracking control points between tactile images. Considering a point of interest, p, in the image of the reference configuration. It is desired to determine q, the point to which it has moved in the image of the deformed configuration. Since the contacted object is a 3-D object and the tactile images are provided in 2-D, the 3-D surface image has been reconstructed from 2-D image using "Shape from Shading" method [22]. The control points are extracted from the reconstructed 3-D image.

The non-rigid pattern matching algorithm we propose, uses iterative algorithm to find appropriate correspondence and transformation function between control points. The displacement of the object deformation is obtained from the transformation function. This displacement function is used to estimate the strain information. The algorithm is described in more detail below.

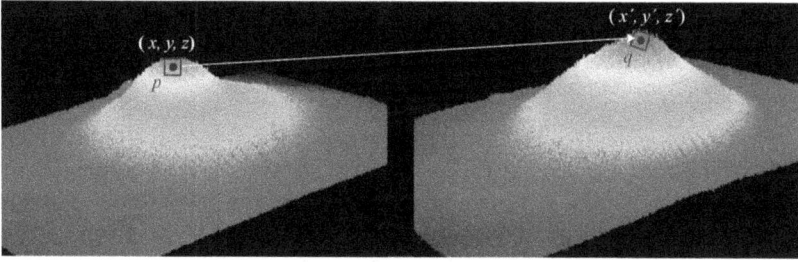

Fig. 5.6. The estimation of displacements of the contacted object deformation by tracking control points on the surface of the tactile image.

If we consider the first 3-D tactile image as a reference image and the second 3-D tactile image with different loading ratio, as the target image, then we can use the pattern matching method to obtain the strain information. From the surface of each 3-D tactile image, a random number of control points are extracted. Let $P = \{p_1, p_2, ..., p_I\}$ be a set of points in the model and $Q = \{q_1, q_2, ..., q_J\}$ be a set of points in the target. For a given point, $p_i \in P$, one can select neighboring points $N_a(p_i)$, $a = 1, 2, ..., A$, which reside in the circle centered at p_i. We set the radius of a circle as the median value of all Euclidean distances between point pairs in P. Similarly, for a point, $q_j \in Q$, adjacent points are $N_b(q_j)$, $b = 1, 2, ..., B$. In this chapter, the pattern matching problem is formulated as a graph matching problem. Each point is a node of a graph, and a given point and its adjacent point constitute the edges of the graph. The problem then is to maximize the number of matched edges between two graphs. For this purpose, we determine the fuzzy correspondence matrix M. Each entry of M has continuous value between $[0,1]$ that indicates the weight of the correspondence between p_i and q_j. The optimal match $\hat{M}$ is found by maximizing the energy function as follows:

$$\hat{M} = \arg\max_{M} C(P, Q, M), \qquad (5.13)$$

where

$$C(P, Q, M) = \sum_{i=1}^{I} \sum_{b=1}^{B} \sum_{j=1}^{J} \sum_{a=1}^{A} M_{p_i q_j} M_{N_a(p_i) N_b(q_j)}. \qquad (5.14)$$

The above equations are subject to the following constraints $\sum_{j=1}^{J+1} M_{p_i q_j} = 1$

for $i = 1, 2, ..., I$ and $\sum_{i=1}^{I+1} M_{p_i q_j} = 1$ for $j = 1, 2, ..., J$.

In the following section, we discuss how the optimal correspondence and transformation function between control points are obtained. For this purpose, the algorithm uses an iterated estimation framework to find appropriate correspondence and transformation.

5.4.2.1. Point Correspondence

Initially, each point $p_i \in P$ is assigned with a set of matching probability based on the shape context distance [22]. After the initial probability assignment, the relaxation labeling process updates the matching probability. The relaxation labeling is an iterative procedure that reduces local ambiguities and achieves global consistency by exploiting contextual information. The process is to assign a matching probability that maximizes $C(P, Q, M)$ under the relaxed condition of $M_{s_i t_j} \in [0,1]$.

At the end of the relaxation labeling process, it is expected that each point will have one unambiguous matching probability. The determination of the compatibility coefficients is crucial because the performance of the relaxation labeling process depends on them. We propose new compatibility coefficient that quantifies the degree of agreement between the hypothesis that p_i matches to q_j and $\mathcal{N}_a(p_i)$ matches to $\mathcal{N}_b(q_j)$.

In the non-rigid degradation of point sets, we note that a point set is usually distorted; however, the neighboring structure of a point is generally preserved due to physical constraints. The displacement of a point and its adjacent point between two point sets constrain one another. Thus, if the distance and angle of a point pair $(p_i, \mathcal{N}_a(p_i))$ in the reference image and its corresponding point pair $(q_j, \mathcal{N}_b(q_j))$ in the target image are similar, we say that they have high correlation. This is further strengthened if a point pair $(p_i, \mathcal{N}_a(p_i))$ in the model shape is closer to each other. To quantify this knowledge, we introduce the similarity constraint α, β as well as the spatial smoothness constraint γ.

The first constraint is the similarity that is related to the differences between the distances and angles of $(p_i, \mathcal{N}_a(p_i))$ and $(q_j, \mathcal{N}_b(q_j))$. This first constraint imposes that if $(p_i, \mathcal{N}_a(p_i))$ has smaller distance and angle differences with $(q_j, \mathcal{N}_b(q_j))$, then they are more compatible. The disparities between $(p_i, \mathcal{N}_a(p_i))$ and $(q_j, \mathcal{N}_b(q_j))$ are defined as follows.

$$\alpha(p_i, \mathcal{N}_a(p_i); q_j, \mathcal{N}_b(q_j)) =$$
$$= \left(1 - \left| (d_i(p_i, \mathcal{N}_a(p_i)) - d_j(q_j, \mathcal{N}_b(q_j))) / \max_{i,j}\{d_i(p_i, \mathcal{N}_a(p_i)), d_j(q_j, \mathcal{N}_b(q_j))\} \right| \right),$$

$$(5.15)$$

$$\beta(p_i, \mathcal{N}_a(p_i); q_j, \mathcal{N}_b(q_j)) =$$
$$= \left(1 - \left| (l_i(p_i, \mathcal{N}_a(p_i)) - l_j(q_j, \mathcal{N}_b(q_j))) / \max_{i,j}\{l_i(p_i, \mathcal{N}_a(p_i)), l_j(q_j, \mathcal{N}_b(q_j))\} \right| \right),$$

$$(5.16)$$

The second constraint, spatial smoothness, is measured by the distance between p_i and $\mathcal{N}_a(p_i)$.

$$\gamma(p_i, \mathcal{N}_a(p_i)) = \left(1 - d_i(p_i, \mathcal{N}_a(p_i)) / \max_{i}\{d(p_i, \mathcal{N}_a(p_i))\}\right), \qquad (5.17)$$

where $\max_{i}\{d(p_i, \mathcal{N}_a(p_i))\}$ is the longest edge of point-adjacent point pairs. We define a total compatibility coefficient by

$$r_{p_i q_j}(\mathcal{N}_a(p_i), \mathcal{N}_b(q_j)) =$$
$$= \alpha(p_i, \mathcal{N}_a(p_i); q_j, \mathcal{N}_b(q_j)) \cdot \beta(p_i, \mathcal{N}_a(p_i); q_j, \mathcal{N}_b(q_j)) \cdot \gamma(p_i, \mathcal{N}_a(p_i)).$$

$$(5.18)$$

The support function $S_{p_i q_j}^k$ in the k-th iteration is given by

$$S_{p_i q_j}^k = \sum_{i=1}^{I}\sum_{j=1}^{J} r_{p_i q_j}(\mathcal{N}_a(p_i), \mathcal{N}_b(q_j)) M_{\mathcal{N}_a(p_i)\mathcal{N}_b(q_j)}^k$$

$$= \sum_{i=1}^{I}\sum_{j=1}^{J} \alpha(p_i, \mathcal{N}_a(p_i); q_j, \mathcal{N}_b(q_j)) \cdot \beta(p_i, \mathcal{N}_a(p_i); q_j, \mathcal{N}_b(q_j)) \cdot \gamma(p_i, \mathcal{N}_a(p_i)) \cdot$$

$$\cdot M_{\mathcal{N}_a(p_i)\mathcal{N}_b(q_j)}^k.$$

$$(5.19)$$

Finally, the fuzzy correspondence matrix element $M_{p_i q_j}$ in Eq. (5.14) is updated according to

$$M_{p_i q_j}^{k+1} = M_{p_i q_j}^k S_{p_i q_j}^k / \sum_{j=1}^{J} M_{p_i q_j}^k S_{p_i q_j}^k, \qquad (5.20)$$

Traditionally, the sum of the rows (or columns) of the matrix M is used as a constraint in the relaxation labeling process. In this chapter we use the sum of the rows and columns as a two-way constraint. In order to meet these constraints, alternated row and column normalization of the matrix M is performed after each relaxation labeling update. This procedure is known as Sinkhorn normalization, which shows that the procedure always converges to a doubly stochastic matrix [25].

5.4.2.2. Transformation Function

Since the strain is determined by the displacements of the contact object deformation, we estimate a transformation function $T : \Re^3 \to \Re^3$ to find the displacements between tactile images obtained under different loading values. In this study, we use the thin-plate spline (TPS) model, which is used in non-rigid pattern matching method for representing flexible coordinate transformations.

Let v_i denote the target function values at corresponding locations $p_i = (x_i, y_i, z_i)$ in the plane, with $i = 1, 2, ..., n$. In particular, we will set v_i equal to x_i', y_i', z_i' in turn to obtain one continuous transformation for each coordinate. We assume that the locations (x_i, y_i, z_i) are all different and are not collinear. In 3-D interpolation problem, the TPS interpolant $f(x, y, z)$ minimizes the bending energy

$$I_f = \iint_{\Re^3} \int \left[\left(\frac{\partial^2 f}{\partial x^2}\right)^2 + \left(\frac{\partial^2 f}{\partial y^2}\right)^2 + \left(\frac{\partial^2 f}{\partial z^2}\right)^2 + 2\left(\left(\frac{\partial^2 f}{\partial x \partial y}\right)^2 + \left(\frac{\partial^2 f}{\partial x \partial z}\right)^2 + \left(\frac{\partial^2 f}{\partial y \partial z}\right)^2 \right) \right] dxdydz,$$

(5.21)

and the interpolant form is

$$f(x, y, z) = a_1 + a_x x + a_y y + a_z z + \sum_{i=1}^{n} w_i U(\|(x_i, y_i, z_i) - (x, y, z)\|).$$ (5.22)

where a_1, a_x, a_y, a_z is the affine transformation coefficients and w_i is the non-affine deformation coefficients. The kernel function $U(r)$ is defined by $U(r) = r^2 \log r^2$ and $U(0) = 0$ as usual. In order for $f(x, y, z)$ to have square integrable second derivative, we require the boundary condition as $\sum_{i=1}^{n} w_i = 0$ and $\sum_{i=1}^{n} w_i x_i = \sum_{i=1}^{n} w_i y_i = \sum_{i=1}^{n} w_i z_i = 0$.

Together with the interpolation conditions, $f(x_i, y_i, z_i) = v_i$, this yields a
linear system for the TPS coefficients:

$$\begin{pmatrix} K & G \\ G^T & 0 \end{pmatrix} \begin{pmatrix} W \\ A \end{pmatrix} \triangleq L \begin{pmatrix} W \\ A \end{pmatrix} = \begin{pmatrix} V \\ 0 \end{pmatrix} = Y, \qquad (5.23)$$

where $K = \begin{bmatrix} 0 & U(r_{12}) & \cdots & U(r_{1n}) \\ U(r_{21}) & 0 & \cdots & U(r_{2n}) \\ \cdots & \cdots & \cdots & \cdots \\ U(r_{n1}) & U(r_{n2}) & \cdots & 0 \end{bmatrix}_{n \times n}$, $G = \begin{bmatrix} 1 & x_1 & y_1 & z_1 \\ 1 & x_2 & y_2 & z_2 \\ \cdots & \cdots & \cdots & \cdots \\ 1 & x_n & y_n & z_n \end{bmatrix}_{n \times 4}$.

Here $r_{ij} = \| p_i - p_j \|$ is the Euclidean distance between points p_i and p_j.
W and A are column vectors formed from $W = [w_1, w_2, ..., w_n]^T$ and
$A = [a_1, a_x, a_y, a_z]^T$, respectively. $V = [v_1, v_2, ..., v_n]^T$ is an arbitrary n-vector.
We denote the $(n+4) \times (n+4)$ matrix $\begin{pmatrix} K & G \\ G^T & 0 \end{pmatrix}$ by L. Define the vector
$Y = (V \mid 0\ 0\ 0)^T$, then we can find the coefficients of TPS by
$(W \mid a_1\ a_x\ a_y\ a_z)^T = L^{-1} Y$ [26].

In the point matching problem, it is necessary to relax the exact
interpolation by means of regularization. This is accomplished by
minimizing the bending energy as follows.

$$H[f] = \sum_{i=1}^{n} (v_i - f(x_i, y_i, z_i))^2 + \lambda I_f \qquad (5.24)$$

The regularization parameter λ, a positive scalar, controls the amount
of smoothing; the limiting case of $\lambda = 0$ reduces to exact interpolation.
As demonstrated in [27, 28] we can solve for the TPS coefficients in the
regularized case by replacing the matrix K with $K + \lambda I$.

After several relaxation labeling updates, the parameters of TPS
deformation model is estimated from the matched control points. The
estimated parameters are then used to transform the reference image
bringing it as close as possible to the target image. The relaxation
labeling process then starts again between the transformed model set and
the target set. The processes for identifying correspondence and

103

transformation are alternatively iterated until the stopping criterion is met. The resulting function $f(x,y,z) = [f_x(x,y,z), f_y(x,y,z), f_z(x,y,z)]$ is a vector-valued, and this maps each point (x_i, y_i, z_i) to (x_i', y_i', z_i') and is the least bent of all such functions. These vector-valued functions $f(x,y,z)$ are the thin-plate spline mappings.

5.4.3. Elasticity Measurement from Stress and Strain

Now we obtain the elasticity from the tactile images using previous stress and strain estimation method. The final TPS model $f(x,y,z) = [f_x(x,y,z), f_y(x,y,z), f_z(x,y,z)]$ provides a continuous displacement field for performing strain analysis. The nonlinear Lagrangian strain tensor components are then determined from the equations as follows

$$e_{xx} = \frac{\partial f_x(x,y,z)}{\partial x} + \frac{1}{2}\left[\left(\frac{\partial f_x(x,y,z)}{\partial x}\right)^3 + \left(\frac{\partial f_y(x,y,z)}{\partial x}\right)^3 + \left(\frac{\partial f_z(x,y,z)}{\partial x}\right)^3\right], \quad (5.25)$$

$$e_{yy} = \frac{\partial f_y(x,y,z)}{\partial y} + \frac{1}{2}\left[\left(\frac{\partial f_x(x,y,z)}{\partial y}\right)^3 + \left(\frac{\partial f_y(x,y,z)}{\partial y}\right)^3 + \left(\frac{\partial f_z(x,y,z)}{\partial y}\right)^3\right], \quad (5.26)$$

$$e_{zz} = \frac{\partial f_z(x,y,z)}{\partial z} + \frac{1}{2}\left[\left(\frac{\partial f_x(x,y,z)}{\partial z}\right)^3 + \left(\frac{\partial f_y(x,y,z)}{\partial z}\right)^3 + \left(\frac{\partial f_z(x,y,z)}{\partial z}\right)^3\right], \quad (5.27)$$

$$e_{xy} = \frac{1}{2}\left(\frac{\partial f_x(x,y,z)}{\partial y} + \frac{\partial f_y(x,y,z)}{\partial x}\right) + $$
$$+ \frac{1}{2}\left[\left(\frac{\partial f_x(x,y,z)}{\partial x}\right)\left(\frac{\partial f_x(x,y,z)}{\partial y}\right) + \left(\frac{\partial f_y(x,y,z)}{\partial x}\right)\left(\frac{\partial f_y(x,y,z)}{\partial y}\right) + \left(\frac{\partial f_z(x,y,z)}{\partial x}\right)\left(\frac{\partial f_z(x,y,z)}{\partial y}\right)\right],$$
$$(5.28)$$

$$e_{yz} = \frac{1}{2}\left(\frac{\partial f_z(x,y,z)}{\partial y} + \frac{\partial f_y(x,y,z)}{\partial x}\right) + $$
$$+ \frac{1}{2}\left[\left(\frac{\partial f_x(x,y,z)}{\partial y}\right)\left(\frac{\partial f_x(x,y,z)}{\partial z}\right) + \left(\frac{\partial f_y(x,y,z)}{\partial y}\right)\left(\frac{\partial f_y(x,y,z)}{\partial z}\right) + \left(\frac{\partial f_z(x,y,z)}{\partial y}\right)\left(\frac{\partial f_z(x,y,z)}{\partial z}\right)\right],$$
$$(5.29)$$

$$e_{zx} = \frac{1}{2}\left(\frac{\partial f_x(x,y,z)}{\partial z} + \frac{\partial f_z(x,y,z)}{\partial x} \right) +$$

$$+ \frac{1}{2}\left[\left(\frac{\partial f_x(x,y,z)}{\partial x} \right)\left(\frac{\partial f_x(x,y,z)}{\partial z} \right) + \left(\frac{\partial f_y(x,y,z)}{\partial x} \right)\left(\frac{\partial f_y(x,y,z)}{\partial z} \right) + \left(\frac{\partial f_z(x,y,z)}{\partial x} \right)\left(\frac{\partial f_z(x,y,z)}{\partial z} \right) \right].$$

$$(5.30)$$

To determine the elastic property of the contacted object from the uniaxial loading configuration, the strain components are averaged e_{zz} along the x and y directions to yield the average strain, $\overline{e}_{zz}$. Given the applied normal stress S_{zz} acting on the loading surface of the optical waveguide, the elastic modulus E is then determined from

$$E = S_{zz} / \overline{e}_{zz}.,$$

$$(5.31)$$

The elasticity estimation using the non-rigid pattern matching algorithm is based on the NP-hard problem and has similar computational complexity of $O(N^3)$ for matching in $\mathbb{R}^3$. For a Young's modulus calculation using a 105×105 point matching, the algorithm takes about 1.69 seconds on a desktop PC with Core 2 Duo CPU with 2.13 GHz and 2 GB RAM.

5.5. Experiment Results

5.5.1. Normal Force Estimation Experiment

In this section, the relationship between the normal force and the integrated pixel value is established via experiments with a loading machine. The loading machine has a force/torque gauge (Mecmesin, West Sussex, UK) to detect the normal force. This machine is shown in Fig. 5.7. The force gauge has a probe to measure the force from 0 to 50 N with a resolution of 1.0×10^{-3} N. Since the camera is an 8-bit digital imager, each pixel value is between of 0 and 255. A circular tip with 2 mm radius is attached to the force/torque gauge and this is used to contact the sensor. To validate the normal force detection, we start from the initial load of 0 N, then the normal force is increased in a stepwise manner. When the applied force reaches around 2.0 N, the applied normal force is decreased in a stepwise fashion until it returns to 0 N.

The resulting diffused light is captured by the camera, and the corresponding contact force is measured by the force gauge.

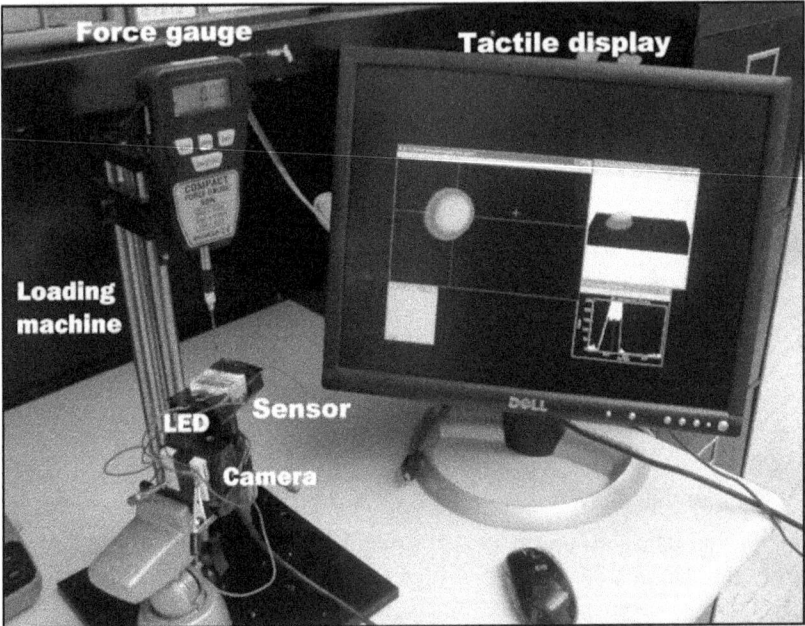

Fig. 5.7. Measurement setup for elasticity-imaging sensor characterization.

Fig. 5.8 shows the pixel value along the contact area's horizontal line passing through the centroid of tactile image. As we expected, the graph is the Gaussian like bell shaped graph and the maximum value is on the centroid of the tactile image. The plot of integrated pixel value change as the applied force changes is shown in Fig. 5.9 (a). The relationship between the integrated pixel value and the applied force is found to be approximately linear as shown in Fig. 5.9 (b). The approximated curve shows a monotone increasing relationship between the normal force and the integrated pixel value of the tactile image. The hysteresis loop is not observed in the graph, indicating that the proposed sensor functions as the precise load cell in the region from 0 to 2 N. Using this approximation, we can approximate the normal forces from the integrated pixel values.

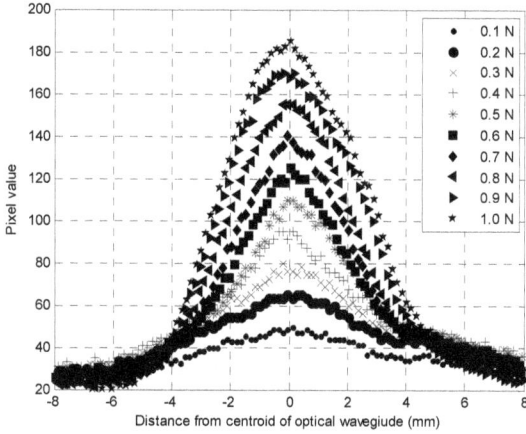

Fig. 5.8. Pixel value along the contact area's horizontal line passing through the centroid of tactile image.

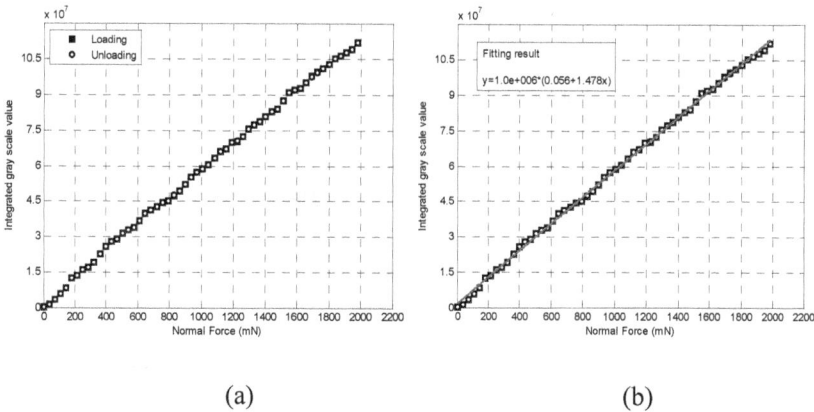

(a)   (b)

Fig. 5.9. The relationship between normal force and integrated gray scale value: (a) Loading and unloading experimental results; (b) the approximated fitting curve.

5.5.2. Strain and Elasticity Measurement Using Soft Polymers

To validate the elasticity measurement using the proposed sensor, Versaflex CL2000X and CL2003X (GLS, McHenry, Illinois) soft polymers with known Young's moduli of 103 kPa and 62 kPa have been used. The objects was 3mm in radius and spherical in shape. The tactile elasticity imaging sensor compressed the polymer samples. The

107

compression ratio was gradually increased. At 0.7 N and 1.2 N applied forces, tactile image has been taken. Fig. 5.10 (a) shows two 2-D tactile images under the 0.7 N and 1.2 N normal forces. In the images, a color scale replaced the original grayscale for better visualization. A purple color indicates grayscale value 0 and a red color indicates grayscale value 255. The two obtained 2-D tactile images were rendered to 3-D images using "shape from shading" method [23]. The 3-D rendered tactile images are represented in Fig. 5.10 (b). The 200 control points were then sampled from the surface of 3-D tactile images. In this the equally spaced control points are extracted automatically. The point correspondence and transformation between control points are iteratively estimated. Fig. 5.11 (a) represents control point distributions from 0.7 N and 1.2 N 3-D tactile images of Vesaflex CL2000X. The final matching result is represented in Fig. 5.11 (b).

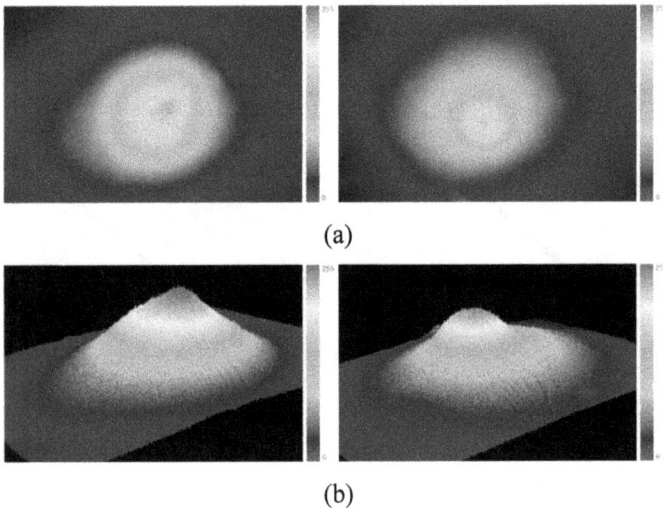

(a)

(b)

Fig. 5.10. 2-D tactile images and 3-D rendered tactile images:
a) 2-D tactile images under 0.7 N (left) and 1.2 N (right) loading value;
(b) 3-D recovery image of 0.7 N (left) and 1.2 N (right).

The TPS transformation functions from the final matching result are used for the elasticity determination. Fig. 5.12 represents the experimental verification. The solid line represents the gold standard of CL2000X and CL2003X moduli, and the square represents measurement values from a tactile elasticity imaging sensor. The errors of the estimated moduli were within 4.23 % for CL2000X and 5.38 % for CL2003X.

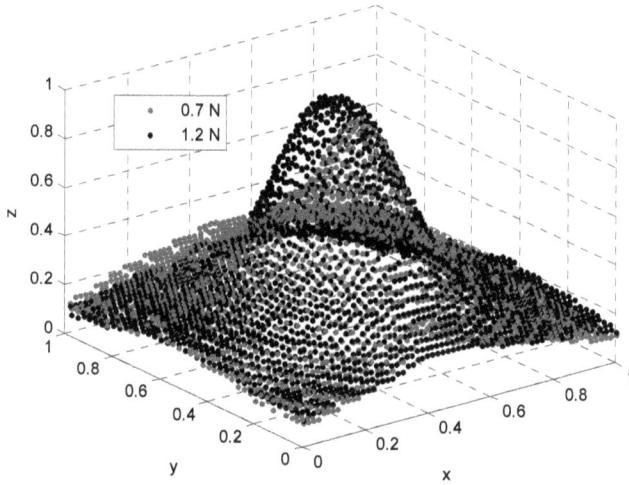

(a)

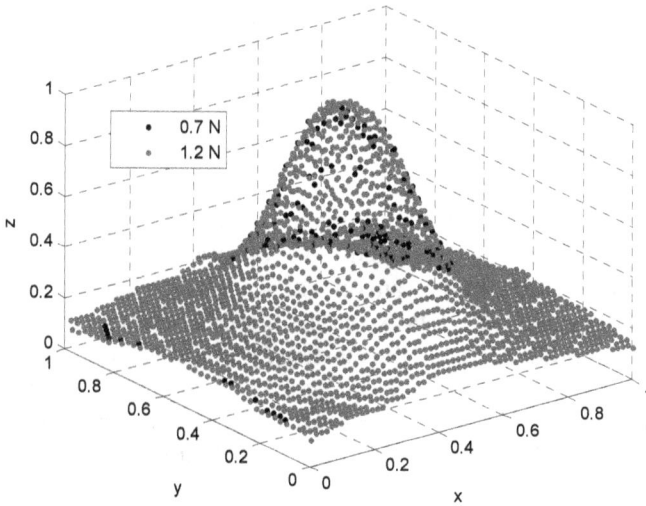

(b)

Fig. 5.11. Non-rigid pattern matching under different tactile image of CL2000X: (a) Control points from 3-D tactile images under the loading values of 0.7 N and 1.2 N to polymer sample, CL2000X; (b) The non-rigid pattern matching results.

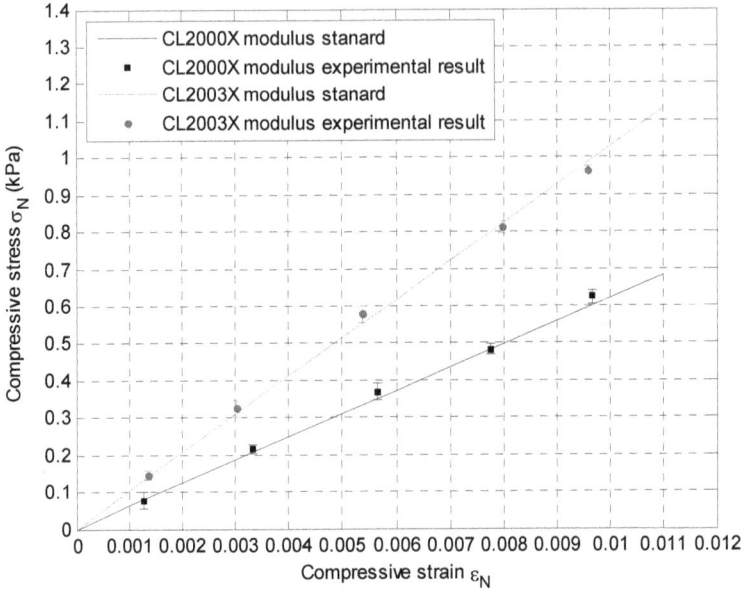

Fig. 5.12. Polymer samples CL2000X and CL2003X moduli measurements using tactile elasticity imaging sensor.

5.6. Conclusions

In this chapter a tactile elasticity imaging sensor using the total internal reflection principle is designed and experimentally evaluated. To increase the sensing range, an optical waveguide consisting of three different densities of PDMS with different elastic modulus was fabricated. In order to obtain the elasticity of the sensed object, the strain is estimated by a non-rigid pattern matching technique. The performance of the proposed sensor is experimentally verified. The results show that the elastic moduli are estimated within 5.38 % of the true value.

Acknowledgments

This research was supported by the Bisa Research Grant of Keimyung University in 2013.

References

[1]. Y. C. Fung, Biomecanics: Mechanical properties of living tissues, *Springer-Verlag*, New York, 1993.

[2]. T. A. Krouskop, T. M. Wheeler, F. Kallel, B. S. Garra and T. Hall, Elastic moduli of breast and prostate tissues under compression, *Ultrason. Imaging*, Vol. 20, 1998, pp. 260-274.

[3]. H. Shojaku, H. Seto, H. Iwai, S. Kitazawa, W. Fukushima, and K. Saito, Detection of incidental breast tumors by noncontrast spiral computed tomography of the chest, *Radiat. Med.*, Vol. 26, 2008, pp. 362-367.

[4]. H. Degani, V. Gusis, D. Weinstein, S. Fields, and S. Strano, Mapping pathophysiological features of breast tumors by MRI at high spatial resolution, *Nat. Mad.*, Vol. 3, 1997, pp. 780-782.

[5]. C. R. Gentle, Mammobarography: a possible method of mass breast screening, *J. Biomech. Eng.*, Vol. 10, 1998, pp. 124-126.

[6]. S. Omata and Y. Terunuma, New tactile sensor like human hand and its applications, *Sens Actuators A*, Vol. 35, 1992, pp. 9-15.

[7]. M. F. Barsky, D. K. Lindner, and R. O. Claus, Robot gripper control system using PVDF piezoelectric sensors, *IEEE Trans. Ultrason., Feroelect., Freq. Contr.*, Vol. UFFC-36, pp. 129-133, Jan 1989.

[8]. K. Motoo, F. Arai, and T. Fukuda, Piezoelectric vibration-type tactile sensor using elasticity and viscosity change of structure, *IEEE Sensors J.*, Vol. 7, 2007, pp. 1044-1051.

[9]. L. Liu, X. Zheng, and Z. Li, An array tactile sensor with piezoresistive single crystal silicon diaphragm, *Sens. Actuators A*, Vol. 32, 1993, pp. 193-196.

[10]. B. J. Kane, A high resolution traction stress sensor array for use in robotic tactile determination, Ph.D. Dissertation, *Stanford Univ.*, Stanford, CA 1999.

[11]. H. Morimura, S. Shigematsu, and K. Machinda, A novel sensor cell architecture and sensing circuit scheme for capacitive fingerprint sensors, *IEEE J. Solid-State Circuits*, Vol. 35, 2000, pp. 724-731.

[12]. D. J. van den Heever, K. Schreve, and C. Sheffer, Tactile sensing using force sensing resistors and a super-resolution algorithm, *IEEE Sensors J.*, Vol. 9, pp. 29-35, 2009.

[13]. H. Yegingil, W. Y. Shih, and W.-H. Shih, All-electrical indentation shear modulus and elastic modulus measurement using a piezoelectric cantilever with a tip, *J. Appl. Phys.*, Vol. 101, 2007, 054510.

[14]. S. Najarian, J. Dargahi, and V. Mirjalili, Detecting embedded objects using haptics with applications in artificial palpation of tumors, *Sens. Mater.*, Vol. 18, 4, 2006, pp. 215-229.

[15]. M. Ohka, H. Kobayashi, J. Takata, and T. Mitsuya, An experimental optical three-axis tactile sensor featured with hemispherical surface, *J. Adv. Mech. Des., Sys., and Manuf.*, Vol. 2, 2008, pp. 860-873.

[16]. S. Saga, H. Kajimoto, and S. Tachi, High-resolution tactile sensor using the deformation of a reflection image, *Sensor Review*, Vol. 27, 2007, pp. 35-42.

[17]. W. R. Uttal, The Psychology of Sensory Coding, *Harper and Row Publishing Co*, 1973.

[18]. E. Kandel, J. Schwartz, and T. Jessell, Principles of Neural Science, *McGraw-Hill Medical*, 2000.

[19]. G. S. Rajan, G. S. Sur, J. E. Mark, D. W. Schaefer, and G. Beaucage, Preparation and characterization of some unusually transparent poly (dimethylsiloxane) nanocomposites, *Journal of Polymer Science*, Vol. 41, No. 16, 2003, pp. 1897-1901.

[20]. D. A. Chang-Yen, R. K. Eich, and B. K. Gale, A monolithic PDMS waveguide system fabricated using soft-lithography techniques, *Journal of Lightwave Technology*, Vol. 23, No. 6, 2005, pp. 2088-2093.

[21]. K. L. Johnson, Contact Mechanics, *Cambridge University Press*, 1985.

[22]. I. Kato, Y. Kudo, K. Ichimaru, Artificial softness sensing - an automatic apparatus for measuring viscoelasticity, *Mechanism and Machine Theory*, Vol. 12, Issue 1, 1977, pp. 11-26.

[23]. P. Tsai and M. A. Shah, Shape from shading with variable albedo, *Optical Engineering*, Vol. 37, 1998, pp. 1212-1220.

[24]. S. Belongie, J. Malik, and J. Puzicha, Shape matching and object recognition using shape contexts, *IEEE Trans. Pattern Analysis and Machine Intelligence,* Vol. 24, No. 4, Apr. 2002, pp. 509-522.

[25]. R. Sinkhorn, A relationship between arbitrary positive matrices and doubly stochastic matrices, *The Annals of Mathematical Statistics*, Vol. 35, No. 2, 1964, pp. 876-879.

[26]. M. J. D. Powell, A Thin Plate Spline Method for Mapping Curves into Curves in Two Dimensions, *Computational Techniques and Applications*, 1995.

[27]. H. Chui and A. Rangarajan, A new point matching algorithm for non-rigid registration, *Computer Vision and Image Understanding*, Vol. 89, No. 23, 2003, pp. 114-141.

[28]. F.L. Bookstein, Principal warps: Thin-plate Splines and the decomposition of deformations, *IEEE Trans. Pattern Analysis and Machine Intelligence*, Vol. 11, No. 6, Jun. 1989, pp. 567-585.

Chapter 6

Integrated Detection of Landmines using Nuclear and Geophysical Sensors

Mohamed Elkattan, Fouad Soliman, R. M. Megahid,
Aladin Kamel and Hadia El-Hennawy

6.1. Introduction

Landmines have been used during warfare for a long time. Unlike conventional weapons, buried landmines, if not removed, remains a threat. Hence, landmine detection is a very important issue, not only in military operation but also in humanitarian concerns. Research conducted in the areas of landmine detection and classification is voluminous encompassing diverse groups of researches and techniques. Several techniques are introduced for landmine detection like nuclear quadruple resonance [1], ground penetrating radar (GPR) [2], thermal IR cameras [3], and chemical sensors [4]. Current demining efforts are heavily reliant on metal detectors and prodders. In many circumstances, the prodder is the first, and in all cases, the last resort. The advent of nondisturbance fused mines makes prodding a dangerous operation. Mechanical devices such as ploughs, rollers, and flails are usually followed by manual demining to obtain the desired level of clearance. These machines are expensive for developing countries. Dogs are good when they work but can only operate for limited periods and must be acclimatized.

Landmine detection involves dealing with wide variety of mine materials and shapes, different terrain and non-uniformity of clutter. Removing an identified anomaly with all the care and attention given to a landmine, to discover in vain that the effort was directed towards clearing a harmless object, is a time consuming and costly process. Thus, it is expected that the characteristic signature for the presence of the landmine should come from multiple sensors sensing different physical properties of the landmine. A multiple sensor landmine detection system should be able to detect mines, and differentiate them from the surrounding clutter.

In this chapter we present an investigational study to obtain the optimal decision fusion between nuclear and geophysical sensors for landmine detection. The multisensor system will work such that the geophysical tools will be suitable for detecting either the casing or the detonator of the landmine. And to distinguish the landmine from the background clutter, nuclear techniques will be used to detect the explosive materials within the landmine. Electromagnetic and magnetic Geophysical techniques will be used in this study for the landmine casing detection. Neutron backscattering technique will be used as the nuclear technique for explosive detection.

6.2. Nuclear Techniques for Landmine Detection

Nuclear Techniques are used to search for the bulk explosive inside the mine based on the property present in explosives which is not present in natural soil or other nearly materials. Nuclear techniques look at either a return radiation, which is characteristic of explosive components that are infrequently found in soil (e.g. nitrogen or Carbon), or an intensity change of a non characteristic scattered radiation, which is a function of a parameter that differs between soil and explosives. Explosives have a much higher percentage of nitrogen (10 to 40 %) than soils (~0.1). Also, the effective atomic number of explosives is between 5 and 7, which is similar to organic material, while for soil it varies from 11 to 12. The typical mass density of soil varies from 1.0 to 2.5 g/cm^3 while for explosives from 1.6 to 1.8 g/cm^3.Nuclear Techniques for landmine detection can be categorized into:

- Nuclear techniques based on density variation;
- Nuclear techniques based on nitrogen density variation;
- Nuclear techniques based on hydrogen density variation;
- Nuclear techniques based on analysis of material whole elements.

Several nuclear techniques have been suggested for detection of landmines and/or explosives, of which some are based on the use of neutrons. Techniques that can give information on the content of the landmine are thermal neutron analysis (TNA) [5], and pulsed fast neutron analysis (PFNA), or a combination of the two [6]. Also, information on the content of the landmine can be gathered from the energy of neutrons that elastically interact in the hidden explosive, known as elastically backscattered spectrometry (EBS) [7]. Although some of these methods have shown great potential, due to the long acquiring time per hidden object, these techniques have only proven to be useful as a confirmation tool.

Nuclear techniques that are used for the localization of anomalies (landmines) in the soil include X-ray [8]. Gamma ray [9], and neutron back scattering (NBS) imaging. The first two techniques are based on changes of the density in the soil. The main problem with these techniques is to achieve an acceptable penetration depth due to attenuation and multiple scattering. Neutron back scattering technique relies on the thermalization of fast neutrons. Since 1999, different detector systems have been developed to use NBS technique for landmine detection [10].

6.3. Neutron Backscattering

6.3.1. Principles

Neutron Back Scattering (NBS) demining techniques takes advantage of the fact that landmines contain much more hydrogen atoms than the dry sand in which they may buried. Hydrogen in the landmine can be found not only in the explosive chemicals, but also in the casing of the plastic landmines. Fig. 6.1 illustrates the principle of Neutron Backscattering. To detect a landmine, soil is irradiated by fast neutrons. These neutrons lose energy by scattering from atoms in the soil and become thermal neutrons after number of collisions. Then, thermal neutron detector monitors the slowed neutrons coming back from the soil [11].

Thermalization process takes many fewer collisions when scattering from hydrogen as compared to other elements. Thus, concentration of thermal neutrons in hydrogenous regions of the soil will therefore be relatively high. The thermal neutron flux at the surface will show an increase above areas which are relatively rich in hydrogen. Such thermal neutron response may indicate the presence of a landmine.

NBS imaging uses a two-dimensional position sensitive neutron detector to produce two-dimensional neutron distribution or image. The advantage of using a two-dimensional image is that the sensitivity for mine detection is greatly enhanced in comparison to only monitoring the overall count rate. The landmine response is not a real image of the landmine because scattering of the neutrons in the soil blurs the image resulting in hot spots which are more or less has circular shapes. Size of the spot has a week dependence on the landmine size, but strongly dependent on the landmine depth [12]. Deeper mines will give hot spots with larger diameters because neutrons that have been thermalized in the landmine will diffuse over larger distances, not only vertically toward the surface, but also horizontally.

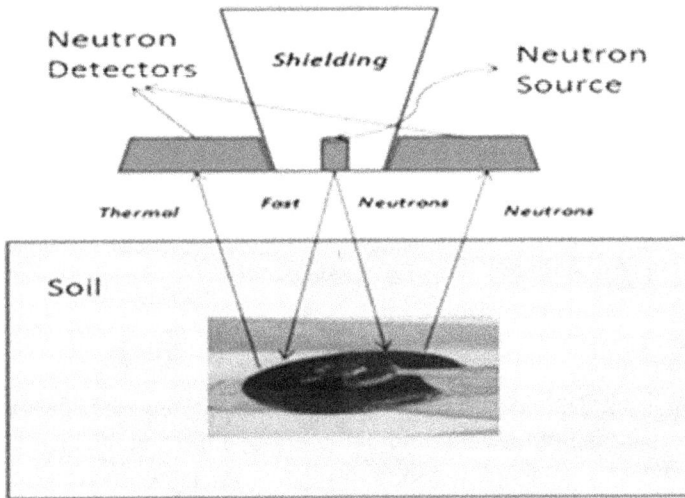

Fig. 6.1. Principle of neutron backscattering technique used in landmine detection.

The main advantage of NBS method is the high speed of operation. Landmines may be found within a second when a sufficiently strong neutron source is used. Additional advantage of NBS based devices over metal detectors is the insensitivities for rocks, stones, and other metal objects, like shells or cartridges of fire arms that used to be in the battle fields. Furthermore, existence of hydrogen in the casing of plastic landmines makes NBS technique an attractive method to detect the fully-metal-free mines, which are used increasingly nowadays.

Since NBS demining technique depends on that the landmines have much higher concentration of hydrogen atoms than the soil, NBS method is sensitive to the soil moisture. Hydrogen content of a landmine is comparable to that of soil with a moisture level of more than 10%, resulting in a loss of contrast between the mine and its surroundings [13]. Moreover, small variations in the distance between the detector and the soil, i.e., the standoff distance, significantly influence the count rate, which in turn will affect the ability of detecting the landmine [14].

116

6.3.2. ESCALAD: Egypt SCAnning LAndmine Detector

A scanning NBS device has been developed at the reactor and neutron physics department of the nuclear research center of the Egyptian Atomic Energy Authority, with the help of Delft University of Technology and supported by IAEA. The detector is based on the Delft DUNBID landmine system [15].

Fig. 6.2 shows The Egypt SCAnning LAnd mine Detector (ESCALAD). It consists of 16 proportional counter tubes with resistive wires, mounted parallel in a flat encasing of 142 cm × 117 cm × 15 cm, perpendicular to the direction of scanning. ^{3}He tubes are sensitive mainly to thermal neutrons through the reaction: ^{3}He + n → p + t + 764 keV. The specifications of the tubes are: length 100 cm, diameter 2.5 cm and ^{3}He pressure 10 bar. The neutron detection efficiency varies from 1 for thermal energy down to 10^{-4} for neutrons in the MeV energy range. The tubes are mounted in two banks of 8 tubes; tube pitch within a bank: 27.8 mm, bank separation: 237.6 mm between the adjacent tubes axes. This separation allows for the placement of radioactive neutron sources or of a neutron generator in the center of the detector [16]. Bank separation is a compromise between two effects:

•The detection sensitivity decreases for an increasing separation distance because of the increase of the path length: source–mine–detector;

• Bull; the 'fast neutron background' increases for smaller separation distances leading to a worse signal-to-background ratio. This type of background is caused by high-energy neutrons, which come straight from the source.

The total sensitive detector area is about 100 cm × 44 cm.

The detector is mounted on a frame allowing it to move or scan horizontally above the ground at a standoff distance of 10 cm. The neutron counts are processed during the movement and are used to build an image of the thermal neutron flux, which is emerging from the ground. The coordinate of a neutron hit in the direction of scanning with respect to a ground related reference frame is determined from the position of the tube being hit in the encasing and the position of the encasing with respect to the ground. The coordinate along the tubes is determined by charge division performed by the electronics. Counts are store according to the measured positions in a 32 × 256 array of pixels with a pixel size of 3.3 cm × 3.4 cm. The first dimension denotes the direction along the tube thus perpendicular to the scan direction; whereas the second one denotes the direction of scanning. The overall background count rate is

117

about 500 c/s for a ^{252}Cf source with an emission rate of 1.6×10^{-5} n/s, and a detector–ground-surface distance of 10 cm. The maximum count rate possible with the current electronics is 2×10^5 c/s, which corresponds to maximum source strength of about 6×10^7 n/s. Each tube in turn spends a time of 3.4/ υ s above a row of pixels, with υ (cm/s) is the scan speed. The total measurement time for the detector tubes to pass completely over a row of pixels, i.e., the pass–over–time is: 16 (3.4/υ).

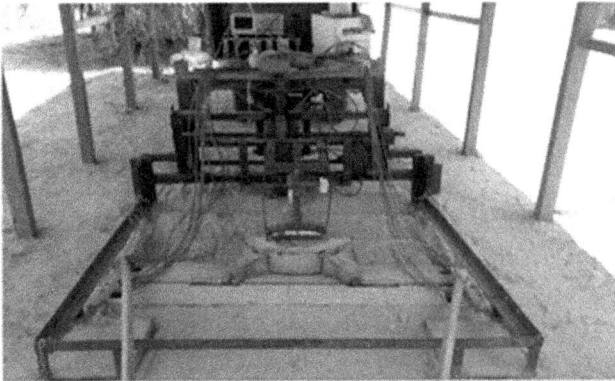

Fig. 6.2. Egypt Scanning Landmine Detector ESCALAD.

6.4. Near Surface Geophysics

Near surface geophysics uses the investigational methods of geophysics to study the nature of the very outermost part of the earth's crust. Many Practitioners argue that a significant percentage of near-surface investigations involve depths of interest less than 10 meters. Near-surface physical properties are highly dynamic due to the various Manhood activities that interact with this part of the earth. This fact leads to technical challenges that are much different than the challenges faced by "traditional" applications of geophysics for regional geologic mapping. The near-surface geophysics application areas can be characterized and distinguished by:

    a.   Having shallow depths of interest;
    b.   Requirements for high resolution, vertically and horizontally;
    c.   The possibility of near-real-time confirmation, verification, or validation of the results.

All the above characteristics are essential in surveys conducted to detect landmines. Thus, there is a great potential for geophysical methods to define subsurface details will a certain level of accuracy. A major challenge for the near-surface geophysics is to exploit the potential of geophysical techniques by being responsive to landmine detection requirements, and congnizant of the capabilities and limitations of each of these techniques [17]. Practitioners of near-surface geophysics often limit data processing and interpretation to a simplistic approach of plotting profiles or contour maps of measured or calculated parameters. The product of this simplistic approach consists of identifying anomalies and recommended locations.

Magnetic and Electromagnetic methods have a prominent place in near-surface geophysics for a number of reasons. First, the sources of interest like man-made objects often have strong signatures. Second, measurements are comparatively simple, rapid, and completely noninvasive. Finally, in near-surface investigations data are often easy to interpret; which make it suitable for real time surveying.

Magnetic technique is one of the best geophysical methods for locating and mapping the distribution of ferro-metallic materials, especially when the signal-to-noise ratio of the magnetic anomalies is high. Measurement of perturbations in the direction and/or strength of the Earth's magnetic field are used to locate underground ferro-magnetic objects. Magnetic data require very few corrections after survey. The main effect that must be compensated is the variation in intensity of the geomagnetic field at the Earth's surface during the course of a day. This "diurnal variation" is due to the part of the Earth's magnetic field that originates in the ionosphere. At any point on the Earth's surface the external field varies during the day as the Earth rotates beneath different parts of the ionosphere. The effect is much greater than the precision with which the field can be measured. The diurnal variation may be corrected by installing a constantly recording magnetometer at a fixed base station within the survey area. The time is noted at which each field measurement is made during the actual survey and the appropriate correction is made from the control record after the survey is done.

Electromagnetic systems measure the electric and magnetic fields generated by time-varying currents in the transmitter loop and nearby conductors [18]. The electromagnetic response of the ground is controlled by the physical properties of the subsurface, the electromagnetic sources and sensors, and measurement parameters, such as geometry, frequency, and measurement time.

119

Phase-component EM measurement methods allow more description of the electrical properties of a buried conductor. These properties affect the phase and relative amplitudes of the in-phase and quadrature "i.e. out of phase" components of the secondary signal relative to the primary. The methods are illustrated by the horizontal loop electromagnetic method "HLEM". In this method, the receiver and transmitter are coupled by a fixed length cable, and kept at a constant separation while the pair is moved along the profile. The cable supplies a direct signal that exactly cancels the primary signal at the receiver, leaving only the secondary field resulted from the suspected object. This is separated into in phase and out of phase components, which are expressed as percentages of the primary field. The in-phase and out-of-phase components are zero far from the conductive body, which enables the outline of a buried object to be charted. The signal rises to a positive peak on either side and falls to a negative peak over the middle of the conductive objects.

6.5. Experimental Setup and Results

6.5.1. Experiment Area and Examined Objects

As a first step toward building an integrated multi-sensor system for detecting the landmines, the proposed geophysical and nuclear techniques was tested in mine field area designed to have similar environmental conditions as the real mine fields in the Egyptian western desert. The test area is sand area with 4 % moisture content, is located near by the laboratories for developing nuclear techniques to detect landmines and illicit materials at the Egyptian Atomic Energy Authority, which is located approximately 60 km northeast of Cairo.

The examined objects were selected to define two categories. The first one is a landmine, either Anti tank or Anti personal, with metal and plastic casing, with different explosive content. The second category represents the objects that considered being "clutter" for the used techniques. Fig. 6.3 indicates the common objects that were scanned by the proposed techniques. Table 6.1 illustrates the characteristic features of the seven objects. In the following, the conducted measurement will be described according to the used method to realize the response of each proposed technique to the examined objects.

120

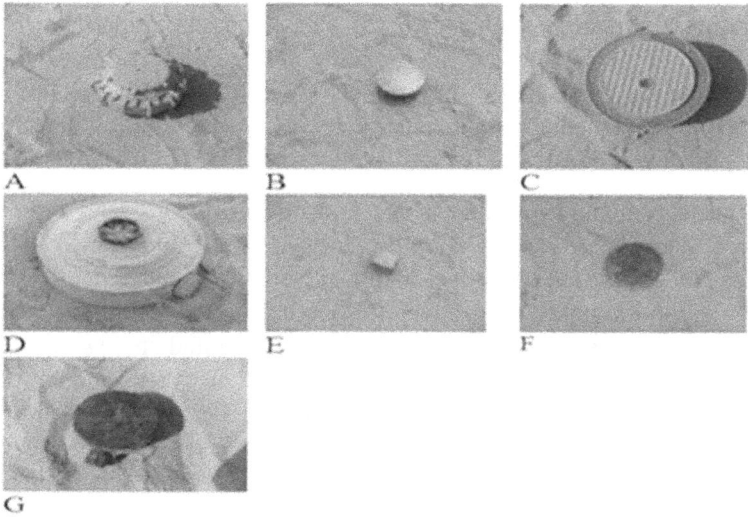

Fig. 6.3. The used objects in the Experiment- their description in Table 6.1.

Table 6.1. Description of Objects.

Object	Type	Size cm	Casing material	Explosive Content, gm
A	VS50 APM	9 × 4.5	Plastic	50 RDX
B	Dummy APM	9 × 4	Bakelite	100 TNT Simulant
C	T-80 ATM	20.4 × 10.8	Plastic	2400 compB
D	T-71 ATM	31.5 × 10	Metal	6000 TNT
E	Cube of wood	10 × 10 x 10	Wood	No Explosive Content
F	Steel Cylinder	10 × 3	Steel	No Explosive Content
G	PMN APM	11.2 × 5.6	Plastic	240 TNT

6.5.2. Electromagnetic and Neutron Backscattering Integration

For EM measurements, HLEM were used to determine the effectiveness of this technique in the detection of the buried landmines. The effective depth of investigation of the HLEM method is controlled by the operating frequency; the geometrical and electrical properties of the target and the transmitter-receiver separation (i.e. coil separation). Since decreasing the frequency of the primary field increases the penetration depth of the induced current, using a suite of frequencies allows a variety of depths to be investigated.

The measurements were done along a profile in the dry sand soil. The HLEM data were collected using two frequencies 3520 Hz and 880 Hz, with a coil separation of 10 m, which make the investigated depth in this study is approximately 5 to 7 meters (one half to two thirds of the coil spacing). The layout of the test profile is illustrated in Fig. 6.4. The station intervals "i.e. measurement points was taken every 1 meter, while a separation distance of 2 meters is selected between two adjacent objects. The seven objects with the same order are used in the EM survey. Furthermore, a piece of metal mounted on each plastic landmine to imitate the landmine "detonator".

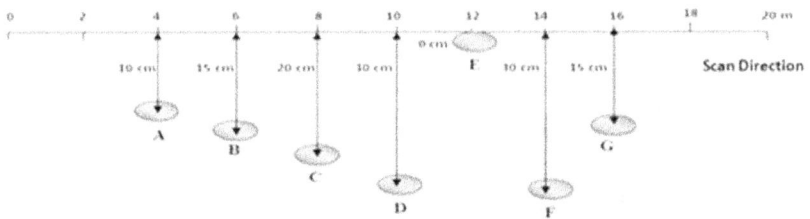

Fig. 6.4. Layout of the test lane.

Also, a metallic disk of dimension 2 m × 2 m is buried at 1m depth and 4 meters away from the center of station 6 "i.e. 6 meters away from the survey starting point". Inspection of the HLEM data for both 3520 Hz, and 880 Hz "see Fig. 6.5." acquired along the 20 meters profile reveals the following:-

- There are no responses for both the plastic and dummy landmines that were buried at stations 4 and 6. This is due to they contain small pieces of metal and buried close to a strong conductor "i.e. metallic disk" which masks the EM signatures of these landmines.

- A strong EM anomaly of a maximum negative peak appeared at station 10 and is associated with both the plastic and metallic landmines buried at stations 8 and 10, as they both detected as a one object body.

- A narrow and strong EM anomaly is observed at station 14 and is associated with the steel cylinder, which has a relatively small dimension compared with the T-71 Anti-tank landmine.

- A relatively weak EM anomaly is related to the plastic landmine that has a metal detonator at station 16.

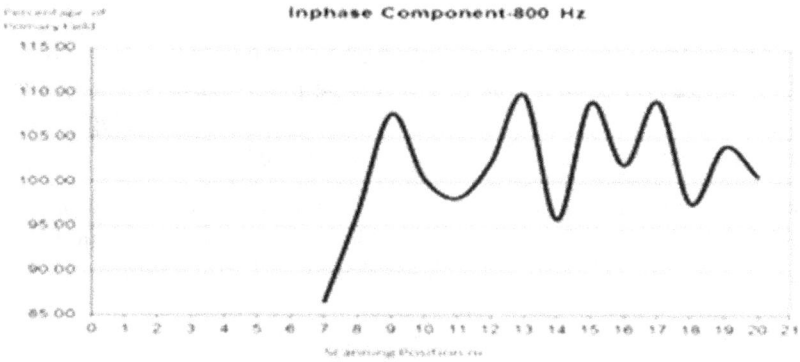

(a)

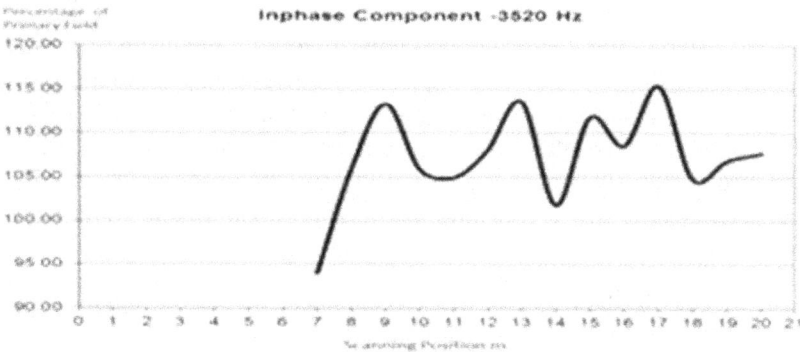

(b)

Fig. 6.5. Horizontal loop electromagnetic data for (a) 800 Hz and (b) 3520 Hz along the test lane with the different buried objects.

For the study of the integration with the Neutron Backscattering Technique, the same test profile was scanned with ESCALAD, which began scanning 2 meters after the HLEM survey starting points. Fig. 6.6 shows the total count resulted from the scanning, which is transformed to 2D Raw Image. Generally, the obtained "Raw Image" of a landmine scan consists of 3 components:

- A signal from the mine "if present";
- A contribution caused by neutrons scatted from the soil;
- A contribution from fast neutrons, which hit the detector without having entered the soil.

The latter two components constitute the background in the image. In the "Raw Image" processing, the row of image pixels perpendicular to the scan direction will be called "pixel raw", while a row of image pixels along the scan direction will be called "pixel line".

The first step in the processing of the raw image is the application of a "fast neutron background" correction. The fast neutron background is an intrinsic system property. It has been measured by placing the detector about 1.5m above the soil and registering the count rate distribution along ach tube. The correction is done by summing the fast neutron measurement along the scan direction and subtracting the result from the pixel rows in the raw image after normalization to the measurement time. The effect of this correction is to remove the central high intensity band and to increase relatively the intensity at the edges of the detector [19].

The next step in the image analysis concerned with the using of imfilter for the enhancement the mine signal over background fluctuation. The process consists simply of moving the center of the filter mask of size 11×5 pixel from point to point in an image. At each point (x, y), the response of the filter at that point is the sum of products of the filter coefficients and the corresponding neighborhood pixels in the area spanned by the filter mask.

It can be shown that ESCALAD detected the entire buried landmine due to its explosive content, except for the dummy antipersonnel mine. Also the signature of the steel cylinder could not be distinguished from the background, which makes ESCALAD a filtration tool for the "metal-non explosive objects" that was detected by the HLEM technique. In the other side, the HLEM technique filtered the "wood cube" which was detected by ESCALAD due to its rich hydrogen content.

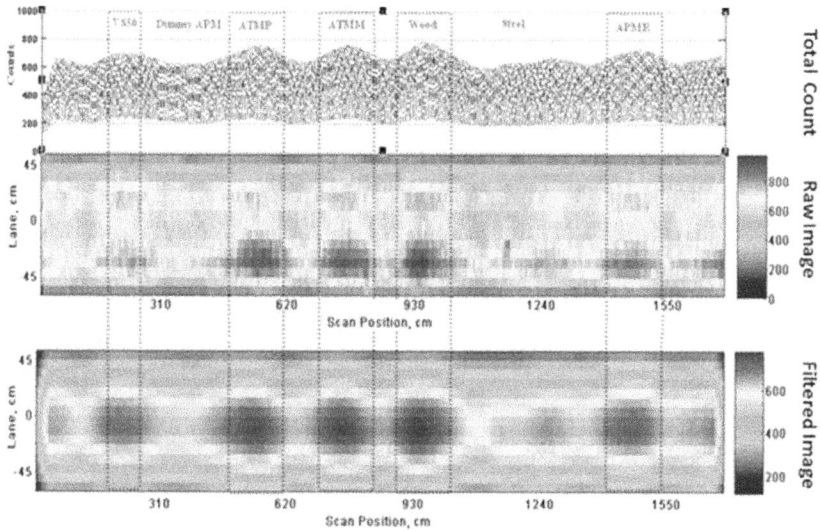

Fig. 6.6. Scan for a test line with different buried objects.

Also, the HLEM technique can solve the relatively limited effective depth of the ESCALAD system, as the effective depth of the HLEM technique depends on the frequency used, ground conductivity and coil separation. However, as decreasing frequency, increasing the penetration depth, but this make the used EM instrument more sensitive to the surrounding buried metal derbies, especially if its relatively big compared to the dimensions of the scanned landmines.

6.5.3. Magnetic and Neutron Backscattering Integration

In this study the magnetic gradient technique has been used for the magnetic measurements. The magnetic gradiometer consists of a pair of magnetometers maintained at a fixed distance from each other at the opposite ends of a rigid vertical bar. During the survey, the difference in outputs of the two magnetometers is recorded. If no anomalous body is present, both magnetometers register the Earth's field equally strongly and the difference in output signals is zero. If a magnetic contrast is present in the subsurface, the magnetometer closest to the object will detect a stronger signal than the more remote instrument and there will be a difference in the combined output signals.

Gradiometers have an advantage of its higher sensitivity with respect to near-surface targets. Moreover, because the gradiometer registers the difference in signals from the individual magnetometers, there is no need to compensate the measurements for diurnal variation, which affect each individual magnetometer equally. Thus, the gradient values do not require any further processing after measurement. One more advantage of using the gradiometer is that it removes to some extent the undesirable influence of disturbing fields.

For studying the magnetometer response over the studied objects, a test lane was conducted in the dry sand area. The same seven objects that were used in the Electromagnetic measurements are used here with the same depths. The layout of the test lane is illustrated in Fig. 6.7. A separation distance of 2 meters is made between each object and the other. For the plastic casing landmines, a small piece of metal was added to each one of them representing the "detonator" of the landmine.

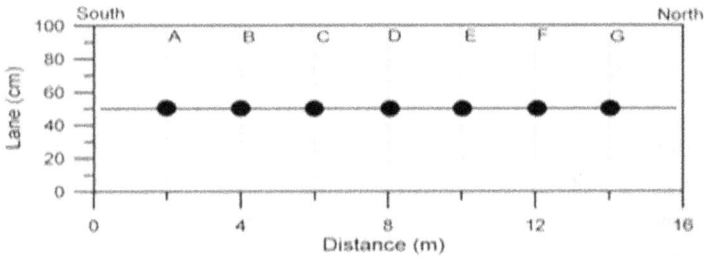

Fig. 6.7. Layout of the conducted test lane. Circles indicate the locations of the objects.

The magnetic instrument used in this experiment is a proton gradiometer with a resolution of 0.01 nT, and a practical measurement rate of once per second. In this test, the instrument was used to measure the vertical gradient, and the distance between the two sensors was 60 cm. The distance between the lower sensor and the ground level was 60 cm making the measurement level from the ground to be 90 cm. The measurements were made at a sampling interval of 30 cm.

Fig. 6.8 shows the recorded vertical magnetic gradient over the buried objects. It can be noticed that many anomalies were recorded but not correlated directly to the landmines. In order to enhance the shape of the magnetic anomalies over their sources, the analytic signal approach was used.

The analytic signal of magnetic anomalies was initially developed as a complex function, and makes use of Hilbert transform properties. The

126

amplitude of the analytic signal is defined as the square root of the squared sum of the vertical and two horizontal derivatives of the magnetic field, where the horizontal and vertical derivatives of the magnetic field are the Hilbert transforms of each other. This transformation produces positive anomaly with peak located nearly above shallow object. The appeal of the method is that analytic signal anomalies can be easily computed, which is suitable for real time detection.

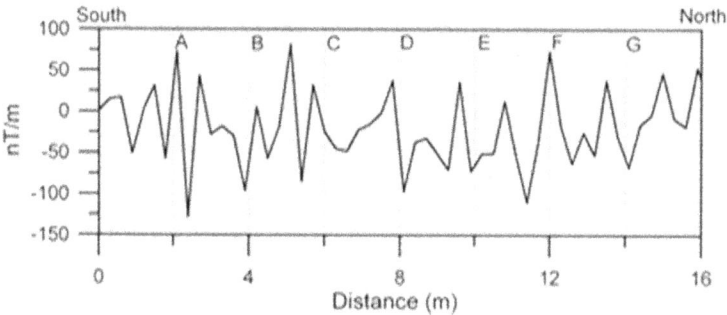

Fig. 6.8. Vertical magnetic gradient profile over the buried objects.

Fig. 6.9 shows the analytic signal of the vertical gradient profile. The analytic signal transformation enhanced the position of the anomalies with respect to the actual location of the objects. While the locations of the peaks over some objects were found slightly shifted, but most of the objects were detected. The non-detected object was the plastic casing AT mine which was buried at 20 cm, and of course the wood sample. Also it can be noticed that some anomalies not related to the objects were also detected like the anomaly between B and C and the anomaly between D and E. These anomalies are most probably related to scattered debris and represent a clutter "unwanted signals.

For that reason, the same test lane was scanned using ESCALAD system. Fig. 6.10 shows the raw image which is formed while scanning. Again, the fast neutron background correction was applied to the raw image, then a proper filter were used to enhance the landmine signal over the background fluctuation. The process consists simply of moving the center of a filter mask of size 3×3 pixel from point to point in the NBS image. At each point (x, y), the response of the filter at that point is the sum of products of the filter coefficients and the corresponding neighborhood pixels in the area spanned by the filter mask.

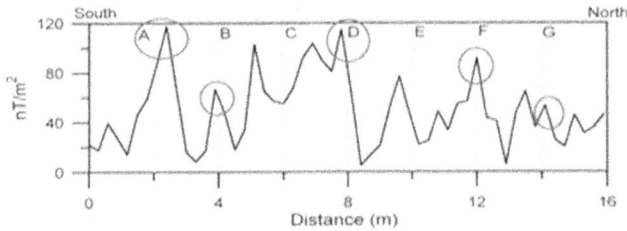

Fig. 6.9. Analytic signal of the vertical gradient magnetic profile.

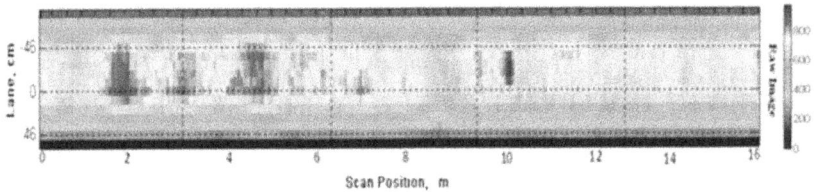

Fig. 6.10. Raw image over the lane with the tested objects by ESCALAD.

Fig. 6.11 shows the filtered image for scanning of the test lane. The obtained image shows that ESCALAD system detected the landmines with different amount of explosives encased in plastic or metal casing. The only object that has not been detected is the piece of steel that was buried at 30 cm depth. Also ESCALAD detected the wood cube, resulting in a hot spot that Look like a landmine signature. Furthermore, more hot spots appeared in the image that are not related to a landmine, but resulted from either a piece of asphalt or sand stone. These unwanted hot spots are a strong source of false alarms for ESCALAD System.

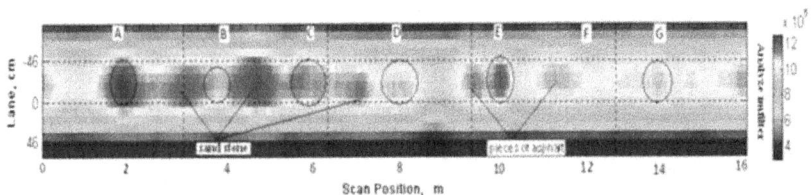

Fig. 6.11. Filtered image over the lane with the tested objects by ESCALAD.

6.6. Multi Sensor Decision Fusion

While the concept of multisensor fusion is not new, it is well known that there is no universal approach for information fusion and that the choice of a particular one strongly depends on the problem itself [20]. The focus here is to apply decision fusion between co-operative sensors to real field data measurements. The decision fusion scheme has the advantage of not requiring knowledge of how the local algorithms are designed and the details of the statistics of the output of the local algorithms.

For the aim of obtaining fusion results as good as possible, and according to the field results, the choice was made to apply the combination between the neutron backscattering and magnetic gradient techniques. The proposed fusion scheme is illustrated in Fig. 6.12. In our algorithm, we first determine the peaks of the triangles in the magnetic analytic signal data, and define the areas in the profile under them as "suspected areas". Then we take the corresponding areas in the neutron backscattering data to determine the possibility of the existence of a "hot spot" in the NBS image.

As mentioned before the hot spot dimensions are variable which is difficult to be identified automatically. We solved this problem, by calculating the angle θ between the "maximum count "and the "minimum count" in the suspected area at the NBS image, see Fig. 6.13. If this angle is more than a certain threshold, the suspected neutron backscattering area considered to have a "Hot Spot" within it. The landmine existence decision is taken if the suspected area has both a peak in the magnetic analytic signal data, and a hot spot in the neutron backscattering data. Otherwise, this location is announced to have no buried landmine.

For the validation of the algorithm, it was tested using the magnetic analytic profile, and the neutron backscattering image resulted from the field test that mentioned earlier in this chapter. Fig. 6.14 shows the data provided by the both techniques, illustrating the triangle peaks, along with the hot spots. By having the analytic signal data as an input, the algorithm detects the existence of 12 peaks, corresponding to 12 suspected areas. Afterwards, the existence of the hot spots in these suspected areas were examined, and the results of the algorithm are shown in Table 6.2.

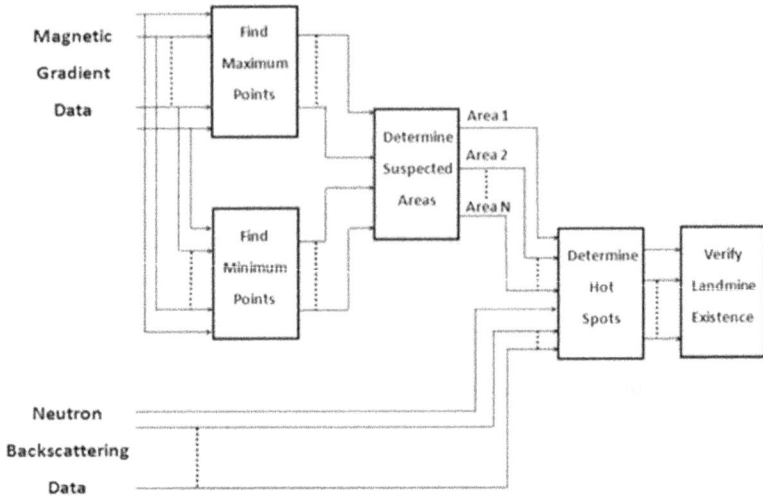

Fig. 6.12. Decision fusion algorithm scheme.

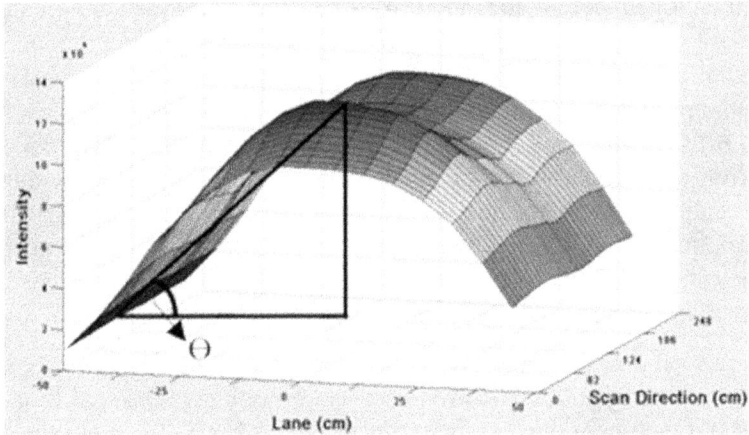

Fig. 6.13. The angle Θ between the maximum count rate and the minimum count rate for the suspected area contains VS50 APM landmine.

It can be noticed that the algorithm detected the first two landmines successfully. Furthermore, the algorithm detected the third landmine, but at this time based on the area under the magnetic anomaly between object B and C, and the existence of a "hot spot" under this area. This can be considered as a false result even there is a landmine there, as the used magnetic anomaly peak is not the real magnetic landmine signature. The same situation happened in the case of the PMN landmine, where the algorithm failed to detect it.

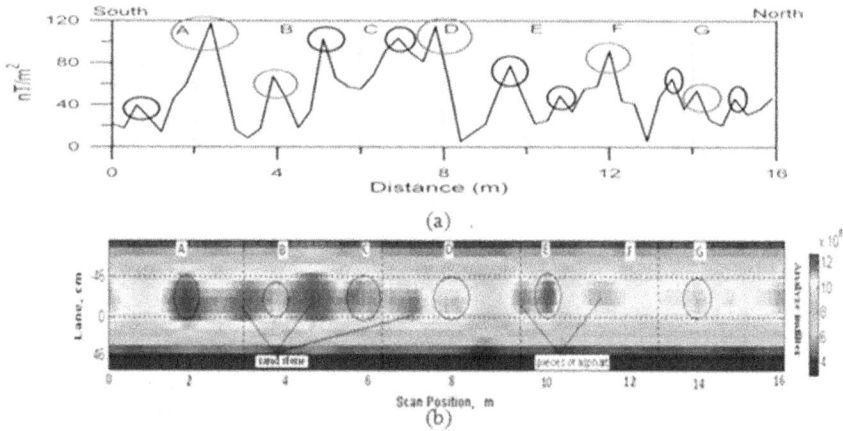

Fig. 6.14. Recorded data provided by (a) magnetic analytic signal, (b) neutron backscattering.

The suspected areas where the cube of wood, and the steel cylinder were buried there, resulted to be no landmine areas, which eliminates the false alarms of both the magnetic and the neutron backscattering techniques, if each one of them stands alone for landmine detection. The rest of the suspected areas that have anomalies related to the scattered debris, resulted to be no landmine areas.

The results of the proposed algorithm concludes to the potential advantage of using an integrated system containing nuclear technique represented on ESCALAD system and conventional technique that depend on the detection of metal parts represented in magnetic gradiometer. However, the fusion algorithm results suggest utilization of more sensors that are able to detect different landmine physical properties. Using more sensors will be very effective in decreasing the false alarms that mislead the proposed fusion algorithm. A confirmation technique like elemental analysis gamma ray sensor can be used for more investigation of such false responses, and would help greatly in the demining process and reduce the risk of hazardous objects. Also, for achieving optimal performance for the fusion algorithm, fine adjustments to the operating thresholds on a case by case basis should be investigated. Such adaptation may not be practical for some military demining operations, but is practical for most humanitarian demining operations.

Table 6.2. Decision fusion algorithm results.

Magnetic Anomaly (Peak)	Nuclear Backscattering (angle Θ)	Fusion Decision	Object Type
1	0.33°	No	No Landmine
2	76.35°	Yes	VS50 APM
3	73.80°	Yes	Dummy APM
4	83.26°	Yes	T-80 ATM
5	52.71°	Yes	No Landmine
6	00.38°	No	T-71 ATM
7	00.89°	No	Cube of wood
8	0.34°	No	No Landmine
9	0.83°	No	Steel Cylinder
10	0.18°	No	No Landmine
11	0.17°	No	PMN APM
12	0.03°	No	No Landmine

Acknowledgements

The authors are greatly indebted to Prof. Assran, S. M. Assran, Exploration Division, Nuclear Materials Authority, for his kind cooperation and constructive comments. The authors are grateful to Prof. Ahmed Salem, Exploration Division, Nuclear Materials Authority, for helping during practical parts of this work.

References

[1]. A. D. Hibb, G. A. Barrall, P. V. Czipott, D. K. Lathrop, Y. K. Lee, E. E. Magnuson, R. Matthews, and S. A. Vierkotter, Landmine detection by nuclear quadruple resonance, in *Proceedings of the SPIE Conference on Detection Remediation Technologies Mines, Minelike Targets III*, Orlando, pp. 522-532, April 1998.

[2]. Polat, B. , and Meinecke, P., Ground penetrating radar imaging of buried metallic objects, in *Proceedings of the IEEE Int. Antennas and Propagation Society Symposium*, Vol. 4, 2001, pp. 264-267.

[3]. M. Lundberg, Infrared land mine detection by parametric modeling, in *Proceedings of the IEEE International Conference on Acoustics, Speech, and Signal Processing. (ICASSP '01)*, Vol. 5, May 2001, pp. 3157 - 3160.

[4]. M. Fisher, C. Cumming, M. Fox, M. La Grone, S. Jacob, D. Reust, M. Rockley, and E. Towers, Sensing ultra-trace concentrations of Landmine chemical signature compounds in the air over landmines using a man-portable chemical sniffer, in *Proceedings of the UXO/COUNTERMINE Forum Conference*, Anaheim, CA, May 2000.

[5]. T. Cousins, T. A. Jones, J. R. Brisson, J. E. McFee, T. J. Jamieson, E. J. Waller, F. J. LeMay, H. Ing, E. T. H. Clifford, E. B. Selkirk, The development of a thermal neutron activation (TNA) system as a confirmatory nonmetallic land mine detector, *J. Radioanal. Nucl. Chem.*, Vol. 235, No. 1-2, 1998, pp. 53-58.

[6]. G. Vourvopoulos, P. C. Womble, and J. Paschal, PELAN: a pulsed neutron probe for UXO, IED and landmine identification, Zagreb, Croatia, IAEA/PS/RC-799, in *Proceedings of the 1st Research Coordination Meeting on Applications of Nuclear Techniques to Anti-Personnel Landmines Identification*, November 1999.

[7]. J. Gsikai, Landmine identification by elastically backscattered Pu-Be neutrons, Zagreb, Croatia, IAEA/PS/RC-799, in *Proceedings of the 1st Research Coordination Meeting on Applications of Nuclear Techniques to Anti-Personnel Landmines Identification*, November 1999.

[8]. S. Shope. G. J. Lockwood, J. C. Wehlburg, M. M. Selph, J. M. Jojola, B. N. Turman, and J. A. Jacobs, Real-time x-ray backscatter imaging of landmines, Zagreb, Croatia, IAEA/PS/RC-799, in *Proceedings of the 1st Research Coordination Meeting on Applications of Nuclear Techniques to Anti-Personnel Landmines Identification*, November 1999.

[9]. L. Zhang, and R. C. Lanza, CAFNA, coded aperture fast neutron analysis for contraband detection: preliminary results, *IEEE Trans. Nucl. Sci.*, Vol. 46, No. 6, 1999, pp. 1913-1915.

[10]. F. D. Brooks, A. Buffler, and S. M. Allie, Detection of plastic land mines by neutron back scattering, in *Proceedings of the 6th Int. Conf. Appl. Neutron Science*, Crete, Greece, January 1999.

[11]. G. Nebbia, DIAMINE (detection and imaging of anti-personnel landmines by neutron backscattering technique), in *Proceedings of the 5th Int. Symp. Technol. Mine Problem*, Monterey, CA, April 2002.

[12]. Victor Bom, A. Mostafa Ali, A. M. Osman, A. M. Abd El-Monem, W. A. Kansouh, R. M. Megahid, and Carel W. E. van Eijk, A feasibility test of land mine detection in a desert environment using neutron back scattering imaging, *IEEE Trans. Nucl. Sci.*, Vol. 53, No. 4, August 2006, pp. 2247-2251.

[13]. J. Obhodas, D. Sudac, K. Nad, V. Valkovic, G. Nebbia, and G. Viesti, The soil moisture and its relevance to the landmine detection by neutron backscattering technique, in *Nucl. Instrum. Methods Phys. Res. B*, Vol. B213, January 2004, pp. 445-451.

[14]. C. Datema, V. R. Bom, and C. W. E. van Eijk, Experimental results and Monte Carlo simulations of a landmine localization device using the neutron back scattering method, *Nucl. Instrum. Methods Phys. Res. A*, Vol. A488, No. 1-2, pp. 441-450, August 2002.

[15]. V. R. Bom, C. W. van Eijk, and M. A. Ali, DUNBID, the delft university neutron backscattering imaging detector, *Appl. Radiat. Isot.*, Vol. 63, No. 5-6, November / December 2005, pp. 559-563.

[16]. Victor R. Bom, R. M. Megahid, A. M. Osman, and M. Musa, ESCALAD, the Egyptian scanning land mine detector, first field test results, in *Proceedings of the 9th Int. Conf. Appl. Nucl. Appl.*, Crete, Greece, June 2008.

[17]. National Research Council, Seeing into the earth, *National Academy Press*, 2000.

[18]. Won, I. J., Keiswetter, K., and Novikova, E., Electromagnetic induction spectroscopy, *Journal of Environmental and Engineering Geophysics,* 3, 1, 1998, pp. 27-40.

[19]. A. M. Osman, M. Musa, S. M. Metwally, S. U. El-Kameesy, and R. M. Megahid, Effect of neutron source geometry on the detection capability of landmines by ESCALAD system, *Arab J. Nucl. Sci. Appl.*, 43, 3, 2010, pp. 165-175.

[20]. Block, I., Information combination operators for data fusion: A comparative review with classification, *IEEE Trans. Syst. Man. Cybern.*, A, Vol. 26, Jan. 1996, pp. 52-67.

Chapter 7

Using Color, Color-Texture and the MLP Neural Network to Select an Optimal Color Space for Human Skin Detection

Hani K. Al-Mohair, Junita Mohamad-Saleh and Shahrel Azmin Suandi

7.1. Introduction

Separation of an image into regions consisting of groups of identical linked pixels is known as image segmentation. The homogeneity of a region is defined by the color, gray levels, and texture, among other features [1]. Skin detection is a good example of image segmentation and is accomplished by classifying the image pixels into skin and non-skin categories using skin color information [2]. Many applications require skin color detection as a primary process, i.e., face detection and recognition [3-5], gesture analysis [6], Internet pornographic image filtering [7], and surveillance systems [8].

Many authors have sought to understand skin color features. Research results showed that skin color has a limited range of hues and is not deeply saturated [9], and therefore, human skin color is clustered within a small area in the color space. Numerous researchers showed that intensity is the main component that differentiates the skin colors of different people [10]. Additionally, conversion of an image into a color space is a useful tool for separating the luminance and chrominance channels, which makes detection of the skin pixels an easy task. [11].

During the past few years, several algorithms have been proposed for skin detection. However, factors such as illumination can make skin color detection a rather difficult task [2]. Texture, which is different from the color, is another significant feature in image segmentation. Texture considers the positional relationships of the pixels and their behavior [12]. In the image-processing field, the use of Artificial Neural Networks have an extensive history of flexibility, more so than the traditional

135

statistical models, and many researchers have reported that Artificial Neural Networks perform well in detection and classification [13-16]. In this study, the Multi-Layer Perceptron Neural Network (MLP NN), an intelligent classifier, is used to evaluate various color spaces to determine the optimal color space for use in skin detection based on the Compaq and ECU databases. Moreover, color features and the combination of color and texture features are highly important factors that have been used with the capabilities of MLP ANN to accomplish this evaluation.

7.2. Color Spaces

Color is a historic source for a broad range of image-processing research fields. However, adjustment of the color space is highly important for image analysis, and thus, colors should be transferred from one space to another to obtain an accurate result [17]. The skin detection procedure is summarized in two steps, i.e., express the image using the perfect color space, and classify the skin pixels based on the assigned skin samples using inference methods [2, 5]. In skin detection, selection of the correct color space is a significant issue. The literature [5, 18] contains several comparisons between different color spaces, but no strict recommendation is given for the appropriate color space for use in skin detection. In addition, many authors do not provide stringent explanations of their color space selections. [17] and certain researchers cannot clarify the conflicting results between their experiments or those of others [19]. Furthermore, certain authors believe that color space selection depends on personal experience rather than scientific experiments [5, 20]. However, most results obtained by researchers showed that different modeling techniques interact quite differently with the change of color space [2, 17]. To address these issues, this chapter presents a comprehensive experiment designed to investigate the best color space for use in skin detection using the MLP ANN. The five color spaces that are investigated and compared are RGB [3], YCbCr [17], YIQ [2], YDbDr [21] and CIE L*a*b [22].

7.3. Texture Features

Based on the fact that the skin region of an image consists of a group(s) of homogenously connected groups of pixels, texture information can be used to describe skin regions. Different texture descriptors, i.e., homogeneity, uniformity, and standard deviation, might be exploited for detection purposes. In this chapter, SD (σ), the skewness and kurtosis of the pixel gray values are used to describe the skin based on the following equations:

$$s = \frac{E(x - \mu)^3}{\sigma^3},$$ (6.1)

The distribution is x, the mean of x is μ, and the standard deviation of x is σ, and E (t) corresponds to the expected value of the quantity t.

$$k = \frac{E(x - \mu)^4}{\sigma^4}$$ (6.2)

7.4. Preparing the Datasets Using Color Features Only

To prepare the dataset, 150 images of skin pixels and 150 images of non-skin pixels were used. The 150 images that contain human skin were downloaded from the "Humanae Project" webpage [23]. The Humanae dataset contains a broad collection of images of different people. In addition, the images of the Humanae dataset are of quite high resolution compared with those of the Compaq dataset [24]. For each image in the skin pixels group, five blocks of 40×40 pixels were manually selected from different areas of the image, as shown in Fig. 7.1.

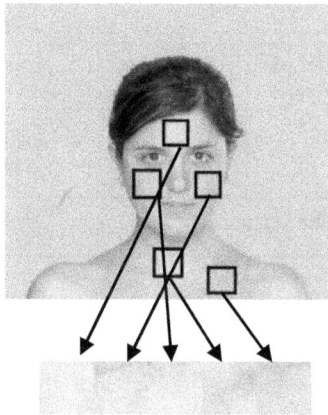

Fig. 7.1. Blocks extracted for training from the Humanae dataset.

137

As a result, the total number of skin pixels collected from 150 images is 1.2 million pixels. A non-skin group of 150 images containing no human skin and downloaded from the internet was used to collect another 1.2 million non-skin pixels. Using the color space transfer equations, each of the resultant datasets (2.4 million pixels) was transformed from the original RGB color space into the targeted colors, i.e., YCbCr, YIQ, YDbDr and CIE L*a*b*. Therefore, five different datasets were obtained and readied for use in training five separate neural networks.

7.5. Combining Color and Texture Features

Because the texture information is collected from training images, the images should be divided into blocks. Therefore, the images are divided into blocks of 4×4 pixels, and the texture information is collected together with color information.

The images (i.e., the image blocks) in the dataset that were previously used to train the ANN based on color information were either pure skin blocks or pure non-skin blocks; this does not affect the process of training because the training process addresses pixels individually. However, if texture information is involved and to render the dataset more realistic, the training data should be modified to contain three different block types: pure skin blocks, pure non-skin blocks, and mixed blocks (blocks containing skin and non-skin areas). This modified dataset is described in the Table 7.1.

Table 7.1. Texture features database distribution.

	No. of blocks	Block size (pixels)	Total pixels	Source
Pure skin blocks	450	40×40	720,000	Humanae database
Pure non-skin blocks	450	40×40	720,000	Collected from the Internet
Mixed blocks	600	40×40	960,000	ECU database
Total size of database			2,400,000	

The resulting dataset contains 1,173,435 skin pixels out of 2,400,000 pixels (equivalent to 48.49 % of the dataset). A scatterplot representing the RGB color space of the skin pixels and non-skin pixels used for training is shown in Fig. 7.2.

The process of collecting training data from the blocks is illustrated in Fig. 7.3. This process can be observed in the resultant matrix, which becomes the training input to the ANN.

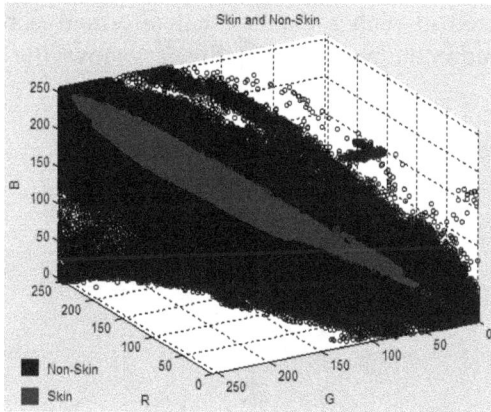

Fig. 7.2. Scatterplot of pixels used for training represented using the RGB color space.

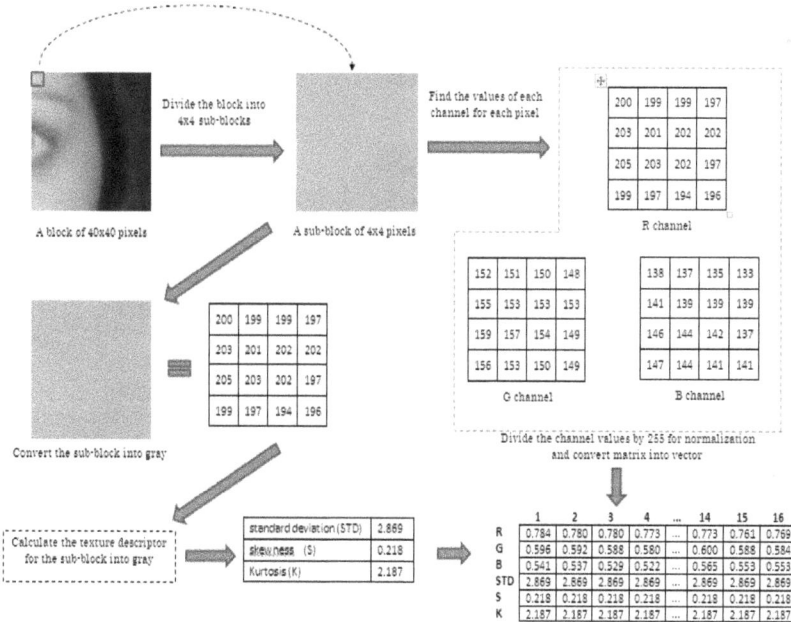

Fig. 7.3. Collecting training data (input) from blocks.

7.6. Training the ANN Using Color Features

The 2.4 million training data pixels are split into three subcategories: training (70 %), validation (15 %) and testing (15 %). In this work, Multi-Layer Perceptron (MLP) ANN is applied as the algorithm. The neurons are interconnected in such a manner that information relevant to I/O mapping is stored in the weights [20]. Fig. 7.4 shows the architecture of the MLP.

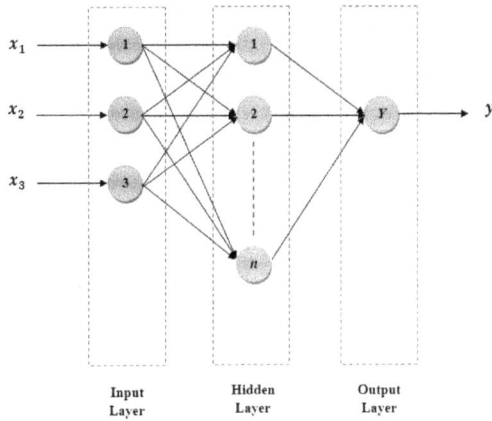

(a)

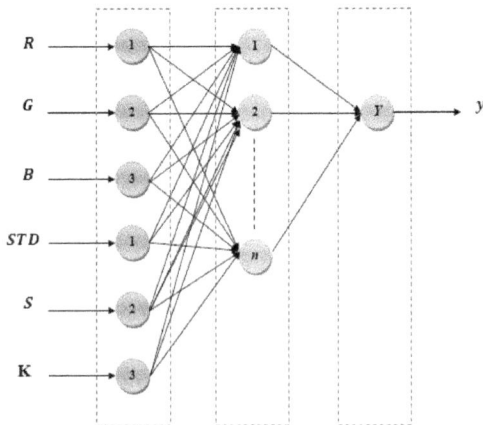

(b)

Fig. 7.4. MLP neural network architecture used for:
(a) Color Features; (b) Color-Texture Features.

The results from the training phase of the different color spaces are shown in Table 7.2. The table illustrates that the best performance is achieved using the YCbCr color space.

Table 7.2. Best validation performance (MSE).

Color Space	Hidden layer neurons	Best validation performance (MSE)
RGB	8	0.099
YCbCr	15	0.098
YDbDr	15	0.099
YIQ	15	0.099
L*a*b	15	0.100

7.7. Training the ANN Using Color–Texture Features

The collected color-texture data from the previous stage are used to train an ANN, as shown in Fig. 7.4. For each configuration, the training process was repeated 20 times in an attempt to determine the best performance of the network in terms of minimum mean square error (MSE). The collected color-texture training data are based on the RGB color space. Four additional color-texture training data arrays based on YCbCr, YDbDr, YIQ, and L*a*b are also generated, which means that five color-texture training datasets are used in total. The five color spaces that were chosen have the best performance if color information only is used for detection [25].

7.8. Testing the ANN Using Color Features

All of the trained neural networks were tested based on two different databases. The first database is a Compaq image database consisting of 100 images (11,340,800 pixels). The second database is the ECU image database containing 800 randomly selected images (212,472,234 pixels) [26]. The Compaq and ECU datasets contain skin pixels from different people under different conditions and were used for testing the trained MLPs, and a range of different thresholds were tested to optimize the results. The actual and predicted classifications produced by the network using a confusion matrix are shown in Table 7.3. Such a matrix is commonly used to evaluate the performance of the MLP and is provided by the following link:

(http://www2.cs.uregina.ca/~dbd/cs831/notes/confusion_matrix/confusion_matrix.html) [27]. The testing process was repeated for all color spaces used in this study. Table 7.3 shows the components of the confusion matrix, and Tables 7.4 and 7.5 show the confusion matrix for the YIQ color space using the Compaq and ECU datasets, respectively.

For comparison purposes, the overall accuracy measures *AC* and *F1-measure* were used for MLP NN performance evaluation. The following Tables 7.6-7.9 illustrate the results.

Table 7.3. Confusion matrix.

AC : Accuracy		Truth Data			
TP : True positive TN : True negative P : Precision		Skin	Non-Skin	Classifica-tion overall	Producer Accuracy (Precision)
Classifier Data	**Skin**	a	b	$a + b$	Precision $P = \dfrac{a}{a+b}$
	Non-Skin	c	d	$c + d$	$\dfrac{d}{c+d}$
	Truth overall	$a + c$	$b + d$	$a+b+c+d$	
	User Accuracy (Recall)	True Positive $TP = \dfrac{a}{a+c}$	True Negative $TN = \dfrac{d}{b+d}$		

Overall Accuracy	$AC = \dfrac{a+d}{a+b+c+d}$

Table 7.4. Confusion matrix for skin color detection using YIQ
with the Compaq Database.

YIQ		Truth Data			
		Skin	Non-Skin	Classifica-tion overall	Producer accuracy (Precision)
Classifier Data	Skin	1,438,614	806,233	2,244,847	64.09 %
	Non-Skin	416,519	8,679,434	9,095,953	95.42 %
	Truth overall	1,855,133	9,485,667	11,340,800	
	User Accuracy (Recall)	77.55 %	91.50 %		

Overall Accuracy	89.22 %

Table 7.5. Confusion matrix for skin color detection using YIQ
with ECU Database.

YIQ		Truth Data			
		Skin	Non-skin	Classification overall	Producer accuracy (Precision)
Classifier Data	Skin	24,114,739	8,608,890	32,723,629	73.69 %
	Non-skin	7,998,576	171,750,029	179,748,605	95.55 %
	Truth overall	32,113,315	180,358,919	212,472,234	
	User Accuracy (Recall)	75.09 %	95.23 %		

Overall Accuracy	92.18 %

Table 7.6. Average accuracy measured by the F1-Measure
with the Compaq Database.

Color Space	Th (Threshold)	TP (%)	FP (%)	Precision (%)	F1-measure (%)
RGB	0.06	80.07	11.02	58.69	67.74
YCbCr	0.04	83.26	11.38	58.86	68.97
YDbDr	0.04	84.44	10.56	60.99	70.82
YIQ	0.06	82.39	9.55	62.80	71.27
L*a*b	0.06	84.08	11.18	59.53	69.71

Table 7.7. Average accuracy measured by the F1-Measure
with the ECU Database.

Color Space	Th (Threshold)	TP (%)	FP (%)	Precision (%)	F1-measure (%)
RGB	0.08	78.27	6.22	69.15	73.43
YCbCr	0.08	78.43	5.63	71.26	74.67
YDbDr	0.06	78.87	5.97	70.19	74.27
YIQ	0.06	80.05	5.98	70.98	75.24
L*a*b	0.08	79.15	6.41	68.73	73.57

Table 7.8. Average accuracy measured by the AC
with the Compaq Database.

Color Space	Th (Threshold)	TP (%)	FP (%)	Precision (%)	AC (%)
RGB	0.06	80.07	11.02	58.69	87.52
YCbCr	0.08	76.36	9.14	62.03	88.49
YDbDr	0.08	79.58	8.89	63.63	89.22
YIQ	0.1	77.55	8.50	64.09	89.22
L*a*b	0.1	78.82	9.51	61.84	88.58

Table 7.9. Average accuracy measured by the AC
with the ECU Database.

Color Space	Th (Threshold)	TP (%)	FP (%)	Precision (%)	AC (%)
RGB	0.18	70.70	4.56	73.41	91.70
YCbCr	0.16	71.94	4.37	74.57	92.05
YDbDr	0.14	72.73	4.52	74.12	92.04
YIQ	0.12	75.09	4.77	73.69	92.18
L*a*b	0.16	72.48	4.80	72.89	91.77

7.9. Testing Using a Combination of Color and Texture Features

Skin detection images using different color spaces for an image from the Compaq dataset and ECU dataset are illustrated in Figs. 7.5 and 7.6, respectively. The best performances (highest *F1* values) are shown in Tables 7.10 and 7.11. A comparison between color-based detection and color-texture-based detection is illustrated in Tables 7.12 and 7.13. The comparison shows small differences between the tested color spaces. However, color-texture detection based on RGB produced the best performance (F1 = 78.21, Accuracy = 93.02) with the ECU database, whereas the color-texture based on L*a*b has the best performance (F1= 74.35, Accuracy= 91.02) when tested on the Compaq database.

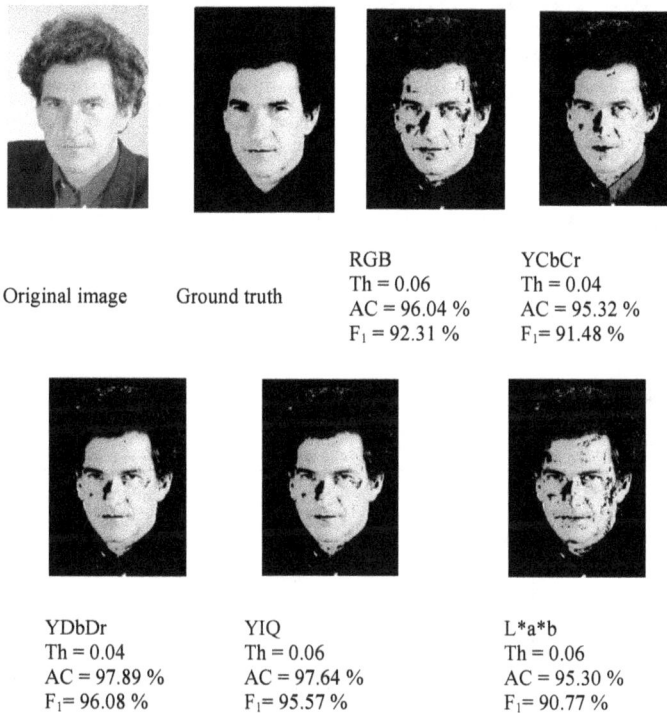

Original image   Ground truth

RGB
Th = 0.06
AC = 96.04 %
F_1 = 92.31 %

YCbCr
Th = 0.04
AC = 95.32 %
F_1= 91.48 %

YDbDr
Th = 0.04
AC = 97.89 %
F_1= 96.08 %

YIQ
Th = 0.06
AC = 97.64 %
F_1= 95.57 %

L*a*b
Th = 0.06
AC = 95.30 %
F_1= 90.77 %

Fig. 7.5. Skin detection using different color spaces for an image from the Compaq Dataset.

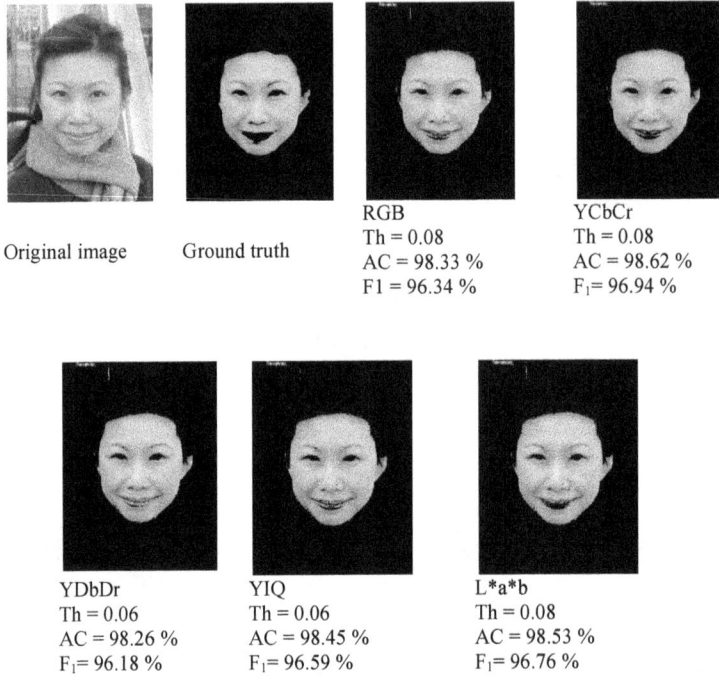

		RGB Th = 0.08 AC = 98.33 % F1 = 96.34 %	YCbCr Th = 0.08 AC = 98.62 % F_1 = 96.94 %
Original image	Ground truth		

YDbDr
Th = 0.06
AC = 98.26 %
F_1= 96.18 %

YIQ
Th = 0.06
AC = 98.45 %
F_1= 96.59 %

L*a*b
Th = 0.08
AC = 98.53 %
F_1= 96.76 %

Fig. 7.6. Skin detection using different color spaces for an image from the ECU Dataset.

Table 7.10. Performance of color-texture-based skin detection using the ECU Database.

Texture-color	No. of neurons in hidden layer	Th	TP (%)	FP (%)	Precision (%)	Accuracy (%)	F1-measure (%)
RGB	7	0.6	81.03	4.79	75.57	93.02	78.21
YCbCr	10	0.61	80.27	4.69	75.77	92.98	77.95
YIQ	15	0.6	81.23	4.94	75.06	92.93	78.03
YDbDr	10	0.53	83.95	5.69	72.96	92.71	78.07
LAB	15	0.61	81.13	4.90	75.19	92.94	78.05

146

Table 7.11. Performance of color-texture-based skin detection
using the Compaq Database.

Color space	Color Only		Color-Texture		Percentage of Enhancement	
	AC (%)	F1-measure (%)	AC (%)	F1-measure (%)	AC (%)	F1-measure (%)
RGB	91.70	73.43	93.02	78.21	1.44	6.51
YCbCr	92.05	74.67	92.98	77.95	1.01	4.39
YIQ	92.18	74.99	92.93	78.03	0.81	4.04
YDbDr	92.04	74.27	92.71	78.07	0.73	5.11
LAB	91.77	73.57	92.94	78.05	1.28	6.08

Table 7.12. A Comparison between color-based detection and
color_texture based skin detection (ECU Database).

Texture-color	No. of neurons in hidden layer	Th	TP (%)	FP (%)	Precision (%)	AC (%)	F1-measure (%)
RGB	7	0.57	82.32	7.94	67.05	90.46	73.91
YCbCr	10	0.63	79.25	6.89	69.30	90.83	73.94
YIQ	15	0.63	79.31	7.07	68.78	90.70	73.67
YDbDr	10	0.61	81.02	7.30	68.53	90.78	74.25
LAB	15	0.68	79.32	6.68	69.98	91.02	74.35

Table 7.13. A Comparison between color-based detection and
color_texture based skin detection (Compaq Database).

Color space	Color Only		Color-Texture		Percentage of Enhancement	
	AC (%)	F1-measure (%)	AC (%)	F1-measure (%)	AC (%)	F1-measure (%)
RGB	87.52	67.74	90.46	73.91	3.36	9.11
YCbCr	88.49	68.97	90.83	73.94	2.65	7.21
YIQ	89.22	71.27	90.70	73.67	1.66	3.36
YDbDr	89.22	70.82	90.78	74.25	1.75	4.84
LAB	88.58	69.71	91.02	74.35	2.76	6.66

7.10. Discussion

From this study, it is obvious that Artificial Neural Networks can play a key role in human skin color classification. Many researchers have compared the impacts of the applied color spaces on the accuracy of human skin detection. Nevertheless, no consensus exists on which color space achieves the highest detection accuracy. Moreover, previous comparisons between color spaces have not been carried out using the MLP ANN, and therefore, in this chapter, different color spaces were used for comparison. The variation in the accuracy using the Compaq dataset is clear; the average accuracy ranges from 67.74 % to 71.27 % measured by the *F1-measure* and from 87.52 % to 89.22 % measured by the *AC*. This variation is lower with the ECU database because the average accuracy ranges from 73.13 % to 75.24 % as measured by the *F1-measure* and from 91.74 % to 92.18 % as measured by the *AC*. However, in both datasets, the YIQ color space produced the most accurate detection as measured by both the *F1-measure* and *AC*. In the second experiment, three statistical features are used to distinguish skin areas from other areas, leading to five texture-color descriptors. The detection accuracy (using the *AC* or *F1-measure*) is nearly the same regardless of the type of color space. However, combining texture with color features increases the capability of human skin color classification, as shown in Tables 7.12 and 7.13. Among the five color spaces used, the L*a*b and YDbDr spaces showed slightly better accuracies as measured by the *F1-measure*. Although 900 images were used for testing, the slight differences between accuracy values do not clearly indicate preference for a certain color space. Nevertheless, the similar accuracy values might be due to the type of statistical features used to describe the texture. This aspect requires extra testing using different types of texture features.

7.11. Conclusion

Several computer vision systems use skin detection as a primary step, and color space information is assumed to have an impact on the accuracy of human skin color classification. In this chapter, we compared five color spaces using the MLP NN algorithm. In addition, certain statistical texture features were used to enhance the accuracy of detection. If color information is used alone, the experimental results showed that the highest detection rate is produced using the YIQ color space, as measured by the AC and *F1-measure*. However, the overall results emphasize that pixel color information alone cannot be used to

148

achieve accurate skin detection, and combining color and texture might lead to much more accurate and efficient skin detection.

Acknowledgment

The authors would like to acknowledge Universiti Sains Malaysia Research Grant Individual (USM-RUI) with No: 1001/PELECT/814092 for the financial support.

References

[1]. C. Liu, P. Chung, Objects extraction algorithm of color image using adaptive forecasting filters created automatically. *International Journal of Innovative Computing, Information and Control*, 7, 10, 2011, pp. 5771-5787.

[2]. P. Kakumanu, S. Makrogiannis, N. Bourbakis, A survey of skin-color modeling and detection methods. *Pattern Recognition*, 40, 3, 2007, pp. 1106-1122.

[3]. J. Chaves-Gonzalez, M. Vega-Rodriguez, J. Gomez-Pulido, J. Sanchez-Perez, Detecting skin in face recognition systems: A colour spaces study. *Digital Signal Processing*, 20, 3, 2010, pp. 806-823.

[4]. Z. Zakaria, N. Isa, S. Suandi, Combining skin colour and neural network for multiface detection in static images, in *Proceedings of the Symposium on Information & Communication Technology (SPICT'09)*, Kuala Lumpur 2009, pp. 147-154.

[5]. A. Abadpour, S. Kasaei, Comprehensive evaluation of the pixel-based skin detection approach for pornography filtering in the internet resources, in *Proceedings of the International Symposium on Telecommunications IST*, 2005, pp. 829-834.

[6]. J. Han, G. Awad, A. Sutherland, Automatic skin segmentation and tracking in sign language recognition, *Computer Vision, IET*, 3, 1, 2009, pp. 24-35.

[7]. J. Lee, Y. Kuo, P. Chung, Detecting nakedness in color images, in Intelligent Multimedia Analysis for Security Applications, *Springer, Berlin*, Heidelberg, 2010, pp. 225-236.

[8]. Z. Zhang, H. Gunes, M. Piccardi, Head detection for video surveillance based on categorical hair and skin colour models, in *Proceedings of the 16th IEEE International Conference on Image Processing (ICIP)*, US, 2009, pp. 1137-1140.

[9]. M. Fleck, D. Forsyth, C. Bregler, Finding naked people, in *Proceedings of the 4th European Conference on Computer Vision*, Cambridge, UK, 1996, pp. 593-602.

[10]. J. Yang, A. Waibel, A real-time face tracker, in *Proceedings of the 3rd IEEE Workshop on Applications of Computer Vision (WACV '96),* Sarasota, Florida, USA, 1996, pp. 142-147.

[11]. L. Chen, J. Zhou, Z. Lid, W. Chen, G. Xiong, A skin detector based on neural network, *IEEE International Conference on Communications, Circuits and Systems,* Vol. 1, 2002, pp. 615-619.

[12]. A. Conci, E. Nunes, Comparing color and texture-based algorithms for human skin detection, in *Proceedings of the Tenth International Conference on Enterprise Information Systems (ICEIS),* Vol. HCI, Barcelona, Spain, 2008.

[13]. L. Yeun, Road sign recognition using affine moment invariant, *Journal of Information and Communication Technology (JICT),* 3, 2, 2004, pp. 59-76.

[14]. R. F. Olanweraju, A. A. Aburas, O. O. Khalifa and A.-H. H. Abdalla, Damagless Digital Watermarking using Complex-Valued Artificial Neural Network, *Journal of Information and Communication Technology,* Vol. 9, 2010.

[15]. N. Isa, M. Mashor, N. Othman, K. Zamli, Application of artificial neural networks in the classification of cervical cells based on the Bethesda system, *Journal of Information and Communication Technology (JICT),* 4, 2005, pp. 77-97.

[16]. P. Saad, N. Jamaludin, S. Kamarudin, A. Bakri, N. Rusli, Rice yield classification using backpropagation network, *Journal of Information and Communication Technology (JICT),* 3, 1, 2004, pp. 67-81.

[17]. C. Liu, A Global Color Transfer scheme between images based on multiple regression analysis, *International Journal of Innovative Computing, Information and Control,* 8, 1A, 2012, pp. 167-186.

[18]. B. Zarit, B. Super, F. Quek, Comparison of five color models in skin pixel classification, in *Proceedings of the International Workshop on Recognition, Analysis, and Tracking of Faces and Gestures in Real-Time Systems,* 1999, pp. 58-63.

[19]. D. Kuiaski, H. Neto, G. Borba, H. Gamba, A Study of the effect of illumination conditions and color spaces on skin segmentation, in *Proceedings of the XXII Brazilian Symposium on Computer Graphics and Image Processing (SIBGRAPI),* Brazil, 2009, pp. 245-252.

[20]. M. Yang, D. Kriegman, N. Ahuja, Detecting faces in images: a survey, *IEEE Transactions on Pattern Analysis and Machine Intelligence,* Vol. 24, No. 1, 2002, pp. 34-58.

[21]. Y. Shi, S. Huifang, Image and video compression for multimedia engineering, *CRC Press,* 2000.

[22]. N. Razmjooy, B. Mousavi, M. Khalilpour, H. Hosseini, Automatic selection and fusion of color spaces for image thresholding, *Signal, Image and Video Processing,* 8, 4, 2012, pp. 603-614.

[23]. Humanæ Project website (http://humanae.tumblr.com/).

[24]. M. Jones, J. Rehg, Statistical color models with application to skin detection, in *Proceedings of the IEEE Computer Society Conference on Computer Vision and Pattern Recognition*, Vol. 1., 1999.

[25]. H. Al-Mohair, J. Saleh, S. Saundi. Impact of color space on human skin color detection using an intelligent system, in Recent Advances in Image, Audio and Signal Processing, *WSEAS,* Budapset. Hungary, 2013.

[26]. S. Phung, A. Bouzerdoum, D. Chai, Skin segmentation using color pixel classification: analysis and comparison. *IEEE Transactions on Pattern Analysis and Machine Intelligence*, Vol. 27, Num. 1, 2005, pp. 148-154.

[27]. Confusion Matrix (http://www2.cs.uregina.ca/~dbd/cs831/notes/confusion_matrix/confusion_matrix. html)

Chapter 8

Design of Narrowband Substrate Integrated Waveguide Bandpass Filters

Ahmed Rhbanou, Seddik Bri, Mohamed Sabbane

8.1. Introduction

Developments in RF communication systems, microwave and wireless communication are characterized by high speed the data transfer and necessitate the dielectric substrates with low losses, the integration is easy with low manufacturing costs. Unfortunately, the traditional technology, either planar or non-planar, is incapable to provide all these characteristics at the same time. In fact, the rectangular waveguides present low insertion losses and good selectivity. However, their production is costly and their integration with other planar circuits requires a specific transition. For planar circuits have a low quality factor, but they have a good compatibility and low cost manufacturing. These constraints led us to use the SIW technology to improve the precision and manufacture of devices with great flexibility in telecommunications systems specific satellites and applications in wireless networks WIMAX and WLAN.

The SIW concept associates the use of planar technology microstrip and the functioning of cavities in which are going to exist volume modes [1-7]. Technically, cavities are included in the substratum and are delimited for the upper and lower faces by the metal plane and for the side faces by rows of metallic holes. This vias have a diameter and spacing small to appear as electric walls [8-15]. However, the change of electrical walls by metallic holes implies that certain modes cannot resonate.

Indeed, the current lines along the side walls of the SIW are vertical; the fundamental mode TE_{10} can propagate efficiently. A microstrip transition is used to interconnect SIW to the planar transmission lines and to match the impedance between a microstrip line and the SIW [16-18]. However,

153

the SIW has been applied successfully to the conception of planar compact components for the microwave and millimeter wave applications, such as filters [19-23]. The advantage of this type of structure is to have a better quality factor and good compatibility. Now the researchers exploit this technology to develop new topologies, to solve the problems of conception for components specially filters. Numerous applications were made on SIW filters for millimeter and sub-millimeter. The results show that the quality factor greater than what can be obtained with planar technology.

On the other, the metamaterials have been one of the popular areas of research in the field of microwaves in the recent past, such that the Split Ring Resonators (SRR) who may exhibit negative permeability and permittivity and hence negative refractive index [24-27]. The complementary split ring resonator (CSRR) is introduced for big use in structures planar [28-35]. The CSRR structure is achieved by etching SRR in the background, which can also realize resonance effect and has found great application in the design of wideband bandpass filter.

In this chapter, the properties of the SIW bandpass filters are carefully studied by using three different methods: iris topology, inductive post topology and CSRRs resonators, starting with theoretical studies of sizing a SIW cavity and comprehensive analysis of transitions in planar lines to achieve perfect impedance matching between SIW technology devices and lines microstrip transmission, then treating the filter synthesis tools and characteristic of CSRR. The simulations of the equivalent circuits were made by the method of moments (MoM) based on commercial software package (ADS). On the other hand the simulations of the structures were made by the finite element method (FEM) based on a commercial software package (HFSS).

8.2. Theoretical Study

8.2.1. Design of the SIW Technology

A substrate integrated waveguide (SIW) is made of metallic via-hole arrays in the substrate between top and bottom metal layers replacing the two metal sidewalls are shown in Fig. 8.1.

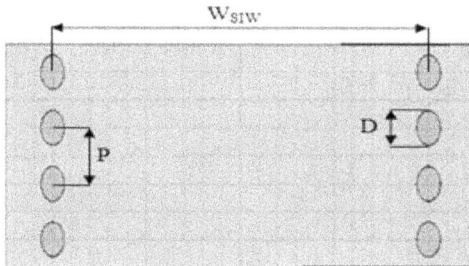

Fig. 8.1. SIW Guide.

The dimension D corresponds to the diameter of vias and P the distance between two adjacent vias center in center. W_{SIW} Is the real distance between the two rows of vias.

The propagation properties in the SIW and in the conventional metallic rectangular waveguide are very similar. In particular, the electromagnetic field distribution is TE_{10} [1-3], the SIW guide is similar to that of a conventional rectangular waveguide filled with the same dielectric of width W_{eff}, with the resonant frequency in the mode TE_{10} is [1-3]:

$$f_{c_{10}} = \frac{c}{2 \times W_{eff} \times \sqrt{\varepsilon_r}} ,$$
(8.1)

The SIW cavity is not an ideal cavity because the electrical walls usually realized by metallic plans in classical volumetric technology, are realized here by rows of vias in SIW (Fig. 8.2).

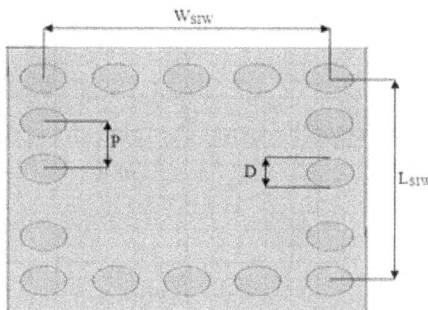

Fig. 8.2. Cavity resonator SIW.

With W_{SIW} and L_{SIW} are the width and the length of cavity SIW. The propagation properties in the SIW and in the conventional metallic rectangular waveguide are very similar, therefore the same for their cavities (Fig. 8.3).

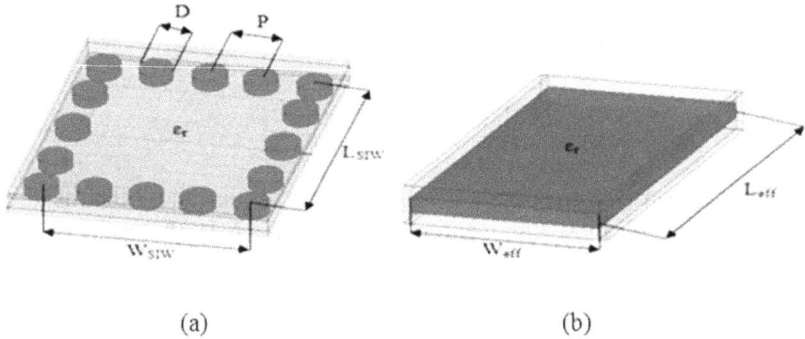

(a)   (b)

Fig. 8.3. (a) Cavity resonator SIW. (b) Rectangular cavity waveguide filled with dielectric.

With W_{eff} and L_{eff} is the length and the width of the rectangular waveguide cavity filled with dielectric, assuming that the dominant resonance mode in the SIW resonator cavity is TE_{101} [4]. Thus, the initial dimensions of the SIW resonator cavity can be determined by the conventional resonant frequency formula of metallic waveguide resonator (f_{c101}) [5, 6]:

$$f_{c_{101}} = \frac{c}{2 \times \sqrt{\varepsilon_r}} \times \sqrt{\left(\frac{1}{W_{eff}}\right)^2 + \left(\frac{1}{L_{eff}}\right)^2} \, , \qquad (8.2)$$

SIW cavity can be designed by using the relations (Eq. 8.2 and Eq. 8.3) provided that $P < \lambda_0 \times (\varepsilon_r/2)^{1/2}$ and $P < 4 \times D$ with ε_r relative permittivity [7-13]:

$$W_{eff} = W_{SIW} - \frac{D^2}{0.95 \times P} \, , \qquad (8.3)$$

$$L_{eff} = L_{SIW} - \frac{D^2}{0.95 \times P} \, , \qquad (8.4)$$

Eq. 8.5 and Eq. 8.6 provide an improvement of the calculation precision provided that P≤2*D and P<λg/5 [14, 15]:

$$W_{eff} = W_{SIW} - 1.08 \times \frac{D^2}{P} + 0.1 \times \frac{D^2}{W_{SIW}}, \qquad (8.5)$$

$$L_{eff} = L_{SIW} - 1.08 \times \frac{D^2}{P} + 0.1 \times \frac{D^2}{L_{SIW}}, \qquad (8.6)$$

8.2.2. Adaptation of SIW Guide

The role of adaptation the microwave device to transmission line is to make a connection with other circuits. In all cases, adaptation must be at most of the order of -15 dB. Generally, the microstrip transitions are very required to combine SIW and microstrip technologies.

The general structure of a microstrip is illustrated in Fig. 8.4. A conducting strip (microstrip line) with a width W_M and a thickness t is on the top of a dielectric substrate that has a relative dielectric constant $\mathcal{E}r$ and a thickness h, and the bottom of the substrate is a ground (conducting) plane.

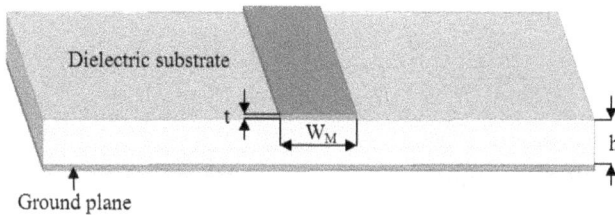

Fig. 8.4. General microstrip structure.

The fields in the microstrip extend within two media—air above and dielectric below—so that the structure is inhomogeneous. Due to this inhomogeneous nature, the microstrip does not support a pure TEM wave. This is because that a pure TEM wave has only transverse components, and its propagation velocity depends only on the material properties, namely the permittivity $\mathcal{E}$ and the permeability μ. However, with the presence of the two guided-wave media (the dielectric substrate and the air), the waves in a microstrip line will have no vanished longitudinal components of electric and magnetic fields, and their propagation velocities will depend not only on the material properties, but also on the physical dimensions of the microstrip.

When the longitudinal components of the fields for the dominant mode of a microstrip line remain very much smaller than the transverse components, they may be neglected. In this case, the dominant mode then behaves like a TEM mode, and the TEM transmission line theory is applicable for the microstrip line as well. This is called the quasi-TEM approximation and it is valid over most of the operating frequency ranges of microstrip.

In the quasi-TEM approximation, a homogeneous dielectric material with an effective dielectric permittivity replaces the inhomogeneous dielectric–air media of microstrip. Transmission characteristics of microstrips are described by two parameters, namely, the effective dielectric constant ε_e and characteristic impedance Z_c. For very thin conductors (i.e., t=0), the closed-form expressions that provide an accuracy better than one percent are given [16] as follows.

For $W_M/h \leq 1$

$$\varepsilon_e = \frac{\varepsilon_r + 1}{2} + \frac{\varepsilon_r - 1}{2} * \left\{ \left(1 + \frac{12 \times h}{W_M} \right)^{\frac{1}{2}} + 0.04 \times \left(1 - \frac{W_M}{h} \right)^2 \right\}, \quad (8.7)$$

$$Z_c = \frac{Z_0}{2 \times \pi \times \sqrt{\varepsilon_e}} \times \ln \left(\frac{8 \times h}{W_M} + \frac{W_M}{4 \times h} \right), \quad (8.8)$$

where $Z_0 = 120 \times \pi$ ohms is the wave impedance in free space.

For $W_M/h \geq 1$

$$\varepsilon_e = \frac{\varepsilon_r + 1}{2} + \frac{\varepsilon_r - 1}{2} * \left(1 + \frac{12 \times h}{W_M} \right)^{-\frac{1}{2}}, \quad (8.9)$$

$$Z_c = \frac{Z_0}{\sqrt{\varepsilon_e}} \times \left(\frac{W_M}{h} + 1.393 + 0.667 \times \ln \left(1.444 + \frac{W_M}{h} \right) \right)^{-1}, \quad (8.10)$$

So far we have not considered the effect of conducting strip thickness t (as referring to Fig. 8.4). The thickness t is usually very small when the microstrip line is realized by conducting thin films; therefore, its effect may quite often be neglected. Nevertheless, its effect on the characteristic impedance and effective dielectric constant may be included [16].

For $W_M/h \leq 1$

$$Z_c = \frac{Z_0}{2 \times \pi \times \sqrt{\varepsilon_e}} \times \ln\left(\frac{8 \times h}{W_e(t)} + \frac{W_e(t)}{4 \times h}\right),$$
(8.11)

For $W_M/h \geq 1$

$$Z_c = \frac{Z_0}{\sqrt{\varepsilon_e}} \times \left(\frac{W_e(t)}{h} + 1.393 + 0.667 \times \ln\left(1.444 + \frac{W_e(t)}{h}\right)\right)^{-1},$$
(8.12)

where

$$\frac{W_e(t)}{h} = \begin{cases} \frac{W_M}{h} + \frac{1.25 \times t}{\pi \times h} \times \left(1 + \ln\left(\frac{4 \times \pi \times W_M}{t}\right)\right) & \text{if } \frac{W_M}{h} \leq \frac{1}{2 \times \pi} \\ \frac{W_M}{h} + \frac{1.25 \times t}{\pi \times h} \times \left(1 + \ln\left(\frac{2 \times h}{t}\right)\right) & \text{if } \frac{W_M}{h} \geq \frac{1}{2 \times \pi} \end{cases}$$
(8.13)

$$\varepsilon_e(t) = \varepsilon_e - \frac{\varepsilon_r - 1}{4.6} \times \frac{t}{h \times \sqrt{\frac{W_M}{h}}},$$
(8.14)

In the above expressions, ε_e is the effective dielectric constant for $t = 0$. It can be observed that the effect of strip thickness on both the characteristic impedance and effective dielectric constant is insignificant for small values of t/h. However, the effect of strip thickness is significant for conductor loss of the microstrip line.

The physical characteristics of the line are determined by (Eq. 8.15, Eq. 8.16 and Eq. 8.17) for given characteristic impedance [17]:

$$\frac{W_M}{h} = \begin{cases} \dfrac{4}{\dfrac{e^{K_1}}{2} - e^{-K_1}} & \text{if } \dfrac{W_M}{h} \leq 2 \\[4mm] \dfrac{\varepsilon_r - 1}{\pi \times \varepsilon_r} \times \left(\ln(K_2 - 1) + 0.39 - \dfrac{0.61}{\varepsilon_r} \right) + \dfrac{2}{\pi}(K_2 - 1 - \ln(2 \times K_2 - 1)) & \text{if } \dfrac{W_M}{h} > 2 \end{cases} \tag{8.15}$$

$$K_1 = \pi \times \sqrt{2(\varepsilon_r + 1)} \times \frac{Z_c}{Z_c} + \frac{\varepsilon_r - 1}{\varepsilon_r + 1} \times \left(0.23 + \frac{0.11}{\varepsilon_r} \right), \tag{8.16}$$

$$K_2 = \frac{\pi \times Z_0}{2 \times Z_c \times \sqrt{\varepsilon_r}}, \tag{8.17}$$

The signal passing through a waveguide requires generally an intermediate transition to link between the planar microstrip circuit technology and the waveguide. This transition has allowed adapting in the fundamental mode impedance of TE_{10} guide on quasi-TEM mode of the microstrip line. A transition must be simple to realize, generate a minimum losses and adaptation must be optimal.

The Transition structures between the planar circuits and conventional rectangular waveguides have been extensively studied and different approaches to adapt impedances were used in microwave.

The transition consists of a type microstrip section that connects a 50 microstrip line and the integrated waveguide; this taper is used to transform the quasi-TEM mode of the microstrip line into the TE_{10} mode in the waveguide. The physical characteristics of microstrip line (the width W_M) and the dimensions (width W_T and length L_T) of a transition are widely detailed in [17]. For our study, using two forms of transition (transition stepped and transition tapered) (Fig. 8.5).

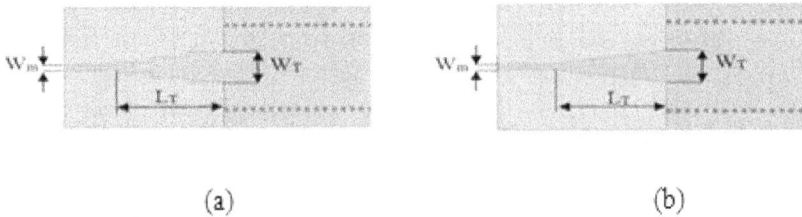

(a)   (b)

Fig. 8.5. (a) SIW Guide with stepped transitions.
(b) SIW Guide with tapered transitions.

160

Once the value of width the microstrip line (W_M) fixed the initial values of W_T and L_T are determined by the following relation [17]:

$$W_T = 0.4 \times (W_{SIW} - D) \quad and \quad \frac{\lambda}{2} \prec L_T \prec \lambda, \qquad (8.18)$$

where λ is the wavelength of the quasi-TEM mode in the microstrip line, in the millimeter frequency range and for a substrate with a permittivity between 2 and 10, the length L_T should be chosen as a multiple of a quarter wavelengths in order to minimize reflection losses. The detailed study was proposed to determine the initial width (W_T) from the following analytical equations [18]:

$$\frac{1}{W_e} = \begin{cases} \dfrac{160}{120 \times \pi} \times \ln\left(\dfrac{8 \times h}{W_T} + \dfrac{W_T}{4 \times h}\right) & if \ \dfrac{W_T}{h} \leq 1 \\[3mm] \dfrac{Z_0}{120 \times \pi \times h} \times \left(\dfrac{W_T}{h} + 1.393 + 0.667 \times \ln\left(\dfrac{W_T}{h} + 1.444\right)\right)^{-1} & if \ \dfrac{W_T}{h} \succ 1 \end{cases} \qquad (8.19)$$

$$\frac{1}{W_e} = \frac{4.38}{W_{SIW}} \times e^{\left(\varepsilon_r + 1 + \dfrac{\varepsilon_r - 1}{2 \times \sqrt{1 + \dfrac{12 \times h}{W_T}}}\right)}, \qquad (8.20)$$

8.2.3. Theoretical Study of SIW Bandpass Filter

The function of filtering can be mathematically described and theoretical tools allow reaching this function from the specifications. The use of these tools constitutes the approach of synthesis, the various steps of which are widely detailed in [19].

The method of synthesis of a bandpass filter consists in determining at first, the low-pass prototype filter equivalent to synthesize as desired. We begin with the choice of topology, which depends on electric properties and characteristics gathered in in the specifications, such as the central frequency, the bandwidth, the insertion losses. Then, you have to choose the type of filter response (Chebyshev, Butterworth, elliptical or pseudo-elliptical).

For our study, we shall use only some elements of the synthesis of a filter of Chebyshev. The latter presents a good rejection and an undulation in the bandpass.

The number of the resonators or the order of the filter is determined by Eq. 8.21 applicable in the case of Chebyshev synthesis [19].

$$n \geq \frac{\cosh^{-1} \sqrt{\dfrac{10^{0.1 L_{As}} - 1}{10^{0.1 L_{Ar}} - 1}}}{\cosh^{-1} \Omega_s} , \qquad (8.21)$$

where n is the order of the filter, L_{As} is the level of out-of-band rejection in the pulsation Ω_s and L_{Ar} is the maximal amplitude of the undulation. $\Omega_s = \Omega$ is the frequency of rejection high ($\omega_s = \omega$), found by the equation of the transformation of frequency [19], whose cut-off frequency is $\Omega_c = 1$ rad/s.

Assume that a lowpass prototype response is to be transformed to a bandpass response having a passband ($\omega_2 - \omega_1$), where ω_1 and ω_2 indicate the passband-edge angular frequency. The required frequency transformation is:

$$\frac{\Omega}{\Omega_c} = \frac{1}{FBW} \times \left(\frac{\omega}{\omega_0} - \frac{\omega_0}{\omega} \right) , \qquad (8.22)$$

with

$$FBW = \frac{\omega_2 - \omega_1}{\omega_0} , \qquad (8.23)$$

$$\omega_0 = \sqrt{\omega_1 \times \omega_2} , \qquad (8.24)$$

where ω_0 denotes the center angular frequency and FBW is defined as the fractional bandwidth.

After determining the order of the filter, the Chebyshev coefficients of the prototype filter are shown by the following [19]:

$$g_0 = 1 , \qquad (8.25)$$

$$g_1 = \frac{2 \times a_1}{\gamma},$$

(8.26)

$$g_k = \frac{4 \times a_{k-1} \times a_k}{b_{k-1} \times g_{k-1}} \qquad k = 2, 3, 4, ..., n,$$

(8.27)

$$a_k = \sin\left[\frac{(2 \times k - 1) \times \pi}{2 \times n}\right] \qquad k = 1, 2, 3, ..., n,$$

(8.28)

$$b_k = \gamma^2 + \sin^2\left(\frac{k \times \pi}{n}\right) \qquad k = 1, 2, 3, ..., n,$$

(8.29)

$$\gamma = \sinh\left(\frac{\beta}{2 \times n}\right),$$

(8.30)

$$\beta = \ln\left(\coth \frac{L_{Ar}}{17.37}\right),$$

(8.31)

$$g_{n+1} = \begin{cases} 1 & \text{for } n \text{ odd} \\ \coth^2\left(\frac{\beta}{4}\right) & \text{for } n \text{ even} \end{cases}$$

(8.32)

The microwave bandpass filters are presented by an equivalent circuit (Fig. 8.6) [19]. This circuit consists of impedance inverters and parallel resonant circuits.

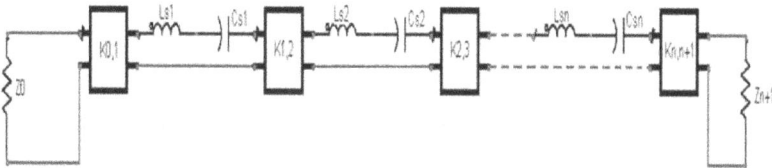

Fig. 8.6. Filter with admittance inverters.

This method of conception is applicable for every type of microwave filters and with various technologies.

From the Chebyshev coefficients, is possible to directly calculate the different elements of the bandpass filter of Fig. 8.6.

The resonators in equivalent circuit are modeled by inductance and capacitance in series [19].

$$L_{si} \times \omega_0 = \frac{1}{C_{si} \times \omega_0} \qquad 1 \leq i \leq n, \qquad (8.33)$$

where ω_0 denotes the center angular frequency, the coupling coefficients between resonators are provided by impedance inverters $K_{i,i+1(0 \leq i \leq n)}$ [19].

$$K_{0,1} = \sqrt{\frac{Z_0 \times FBW \times \omega_0 \times L_{S1}}{\Omega_c \times g_0 \times g_1}}, \qquad (8.34)$$

$$K_{i,i+1} = \frac{FBW \times \omega_0}{\Omega_c} \times \sqrt{\frac{L_{Si} \times L_{S(i+1)}}{g_i \times g_{(i+1)}}} \qquad 1 \leq i \leq n-1, \qquad (8.35)$$

$$K_{n,n+1} = \sqrt{\frac{Z_{n+1} \times FBW \times \omega_0 \times L_{Sn}}{\Omega_c \times g_n \times g_{n+1}}}, \qquad (8.36)$$

where Z_0 is the source impedance, so the filter coupling matrix is presented in following form (Fig. 8.7):

	S	1	2	...	$i-1$	...	n	L
S	0	$K_{0,1}$	0	0	0	0	0	0
1	$K_{0,1}$	0	$K_{1,2}$	0	0	0	0	0
2	0	$K_{1,2}$	0	......	0	0	0	0
...	0	0	......	0	$K_{i,i-1}$	0	0	0
$i-1$	0	0	0	$K_{i,i+1}$	0	......	0	0
...	0	0	0	0	......	0	$K_{n-1,n}$	0
n	0	0	0	0	0	$K_{n-1,n}$	0	$K_{n,n+1}$
L	0	0	0	0	0	0	$K_{n,n+1}$	0

Fig. 8.7. Coupling matrix of the microwave bandpass filter circuit.

The indices S and L correspond to the source and to the load respectively, in other words to the accesses. The indices going from 1 to n correspond to the resonators.

On the other the waveguide filters are formed with resonator distributed elements interconnected by impedance inverters or admittance. The distribution of the electric field in the SIW has characteristics of dispersal similar to the mode of the waveguide. The conception of filter SIW uses the same process the conception of a filter waveguide. The equivalent circuit of the bandpass filter SIW is presented by impedance inverter and phase shifts (Fig. 8.8) [20- 22].

164

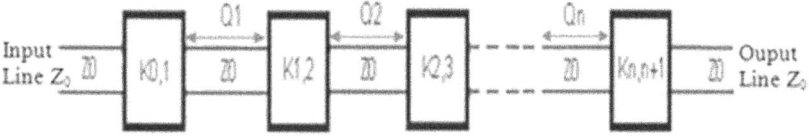

Fig. 8.8. Equivalent circuit of SIW filter.

The impedance inverters $K_{i,i+1(0 \leq i \leq n)}$ are given by the formulas in [23].

$$\frac{K_{0,1}}{Z_0} = \sqrt{\frac{\pi \times \omega_\lambda}{2 \times \Omega_c \times g_0 \times g_1}}, \qquad (8.37)$$

$$\frac{K_{i,i+1}}{Z_0} = \frac{\pi \times \omega_\lambda}{2 \times \Omega_c} \times \sqrt{\frac{1}{g_i \times g_{(i+1)}}} \qquad 1 \leq i \leq n-1, \qquad (8.38)$$

$$\frac{K_{n,n+1}}{Z_0} = \sqrt{\frac{\pi \times \omega_\lambda}{2 \times \Omega_c \times g_n \times g_{n+1}}}, \qquad (8.39)$$

where ω_λ is the fractional bandwidth [23], defined by the guided wave lengths (λ_{g1} and λ_{g2}) for cutoff frequencies low (f_1) and high (f_2) of the bandwidth and the guided wave length (λ_{g0}) of the center frequency f_0:

$$\omega_\lambda = \frac{\lambda_{g1} - \lambda_{g2}}{\lambda_{g0}}, \qquad (8.40)$$

$$\lambda_{gi} = \frac{1}{\sqrt{\left(\dfrac{f_i \times \sqrt{\varepsilon_r}}{c}\right)^2 - \dfrac{1}{\lambda_c^2}}} \qquad i = 0,1,2, \qquad (8.41)$$

In hybrid networks the impedance inverter is a broadband circuit (Fig. 8.9).

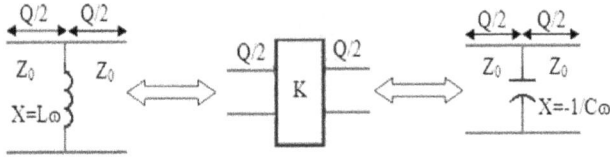

Fig. 8.9. Broadband impedance inverters.

The equivalence relation with the K-inverter is given by the formula in [23].

$$\left| \frac{X_{i,i+1}}{Z_0} \right| = \frac{\dfrac{K_{i,i+1}}{Z_0}}{1 - \left(\dfrac{K_{i,i+1}}{Z_0} \right)^2} \tag{8.42}$$

The electrical lengths of the resonators are given by the formulas in [23].

$$\varphi_i = \pi + \frac{1}{2} \times \left[\theta_{i-1,i} + \theta_{i,i+1} \right] \qquad 1 \le i \le n, \tag{8.43}$$

$$\theta_{i-1,i} = -\tan^{-1}\left(\frac{2 \times X_{i-1,i}}{Z_0} \right) \qquad 1 \le i \le n, \tag{8.44}$$

The topologies of SIW have been extensively studied. For our study we used the topology with iris and topology with circular inductive post.

The SIW filter with iris is influenced by the lengths L_{SIWi} (i=1,2,3...,n) the resonators and also by the coupling, That is an opening in the wall between two adjacent cavities. This type of opening is called iris (the widths W_{SIWi} (i=0,1,2.....n) the resonators) as shown in Fig. 8.10.

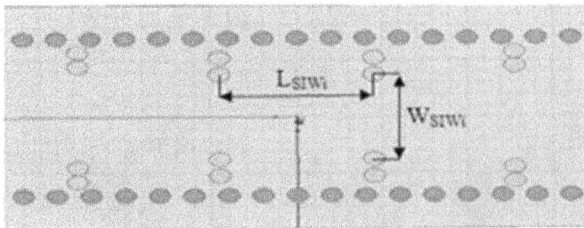

Fig. 8.10. SIW filter with iris.

Iris is represented by a diagram (or circuit) consisting of two equivalent capacitances X_S and self X_P as shown in Fig. 8.11.

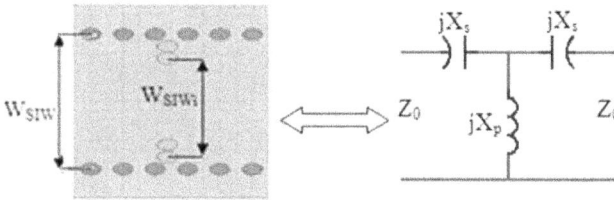

Fig. 8.11. Iris and his equivalent circuit.

In the Fig. 8.11 W_{SIW} is the width of SIW guide, W_{SIWi} is the width of the resonator. When the diameter of plated cylindrical holes (vias) is very small compared to the width of the waveguide, the equivalent circuit becomes equivalent to a parallel inductance ($X_S=0$).

The lengths L_{SIWi} (i=1,2,3...,n) of the resonators are presented by formula in [23]:

$$L_{SIWi} = \frac{\lambda_{g0} \times \varphi_i}{2 \times \pi} \quad 1 \leq i \leq n, \tag{8.45}$$

where φ_i ($1 \leq i \leq n$) are the electrical lengths of the resonators. The widths W_{SIWi} (i=0,1,2.....n) of the resonators in SIW technology are determined by the abacus of the estimated width in rectangular waveguide.

The SIW filter with circular inductive post is influenced by the lengths L_{SIWi} the cavities and by the diameter of circular post d_{SIWi} which delimit the cavities of the filter as shown in Fig. 8.12.

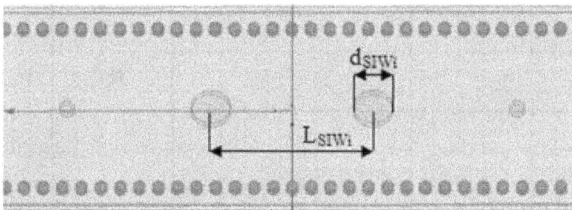

Fig. 8.12. SIW filter with circular inductive post.

Circular post is represented by a diagram (or circuit) consisting of two equivalent capacitances X_S and self X_P as shown in Fig. 8.13:

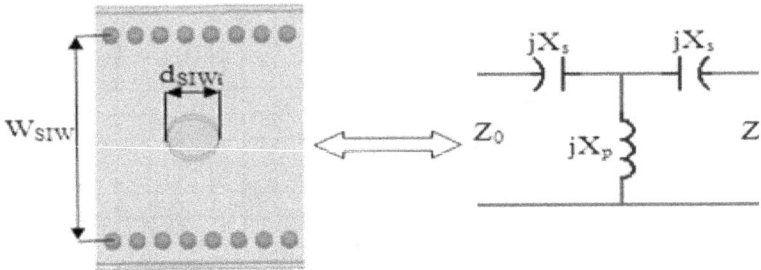

Fig. 8.13. Typical equivalent circuit for circular post.

In the Fig. 8.13 W_{SIW} is the width of the waveguide and d_{SIWi} the diameter of circular post, the lengths L_{SIWi} (i=1,2,3...,n) the cavities SIW inductive post filter are determined by the same method the SIW filter with iris [23]. The diameter of circular posts d_{SIWi} (i=0,1,2,....n) is determined by the abacus of the estimated diameter in rectangular waveguide [23].

8.2.4. Complementary Split Ring Resonators 'CSRR'

The electromagnetic properties of SRRs behave as an LC resonator [24-27]. Split Ring Resonator (SRR) is a well-known sub-wavelength metamaterials structure that exhibits negative values of permeability over a narrow frequency band around it resonance frequency. The resonant frequency is determined from the geometrical parameters of SRR. The SRR can have different types of structures (square, circular, Omega ...) with single ring, double ring or multiple ring SRR cells. The Complementary Split Ring Resonator CSRR is the complementary of SRR [28-32]. The CSRR the rings are etched on a metallic surface and its electric and magnetic properties are interchanged with respect to the SRR. Fig. 8.14 shows the difference between the SRR and the CSRR. In fact, all the conductive part (rings) and the dielectric part of the SRR are respectively replaced by the dielectric and conductive plan of a substrate in the CSRR.

In this work, the square single ring CSRRs (Fig. 8.14 (b)) cells and the square double rings CSRRs cells (Fig. 8.15) are carefully studied for designing the bandwidth filter.

168

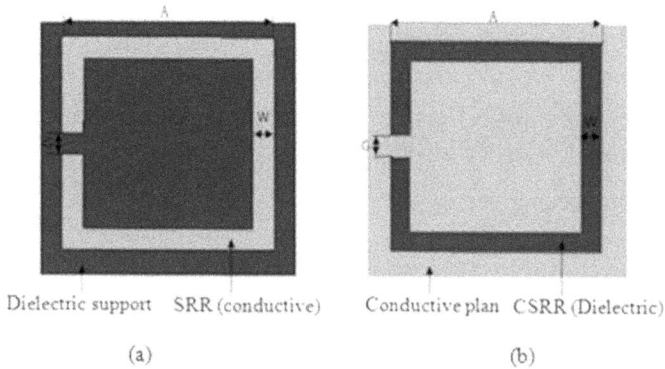

Dielectric support   SRR (conductive)   Conductive plan CSRR (Dielectric)

(a)   (b)

Fig. 8.14. (a) Square-shaped single-ring SRR. (b) Square-shaped single-ring CSRR.

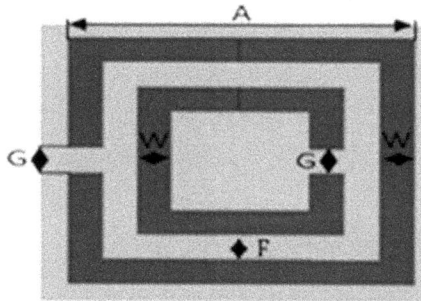

Fig. 8.15. Geometry square CSRR of the double rings.

In the Fig. 8.15 "A" denotes the length of the side of the square, "W" denotes the width of the conductor, "F" denotes the dielectric width between the inner and the outer square and "G" denotes the gap present in the rings. Care is taken that the gap width "G" does not change from the inner ring to the outer ring.

The resonance frequency is obtained by is by using the equivalent circuit analysis method as prescribed in [33]. When a magnetic field is applied perpendicularly to the plane of the ring, the ring begins to conduct and gives rise in current flow. The current flowing through the rings will enable it to act as an inductor and the dielectric gap (F) between the rings will lead to mutual capacitance. Hence the equivalent circuit of the square CSRR with double rings will be a parallel LC resonant circuit in [33]. The resonance frequency is calculated by Eq. 8.46.

$$f = \frac{1}{2 \times \pi \times \sqrt{LC}},$$ (8.46)

The expressions for effective inductance and capacitances can be obtained from [34] as follows:

The capacitance is given by

$$C = 4 \times \left(\frac{\varepsilon_0}{\mu_0}\right) \times L_1,$$ (8.47)

With L_1 present the inductance of the equivalent circuit proposed for square SRR in [35].

$$L_1 = \frac{4.86 \times \mu_0}{2} \times (A - W - F) \times \left[\ln\left(\frac{0.98}{\rho}\right) + 1.84 \times \rho\right],$$ (8.48)

where A, W and F are the notations prescribed in the previous section, ρ is the filling factor of the inductance and is given by

$$\rho = \frac{W + F}{A - W - F},$$ (8.49)

The inductance is given by

$$L = \left(\frac{\mu_0}{4\varepsilon_0}\right) \times C_1,$$ (8.50)

With C_1 present the capacitance of the equivalent circuit proposed for square SRR in [35].

$$C_1 = \left(A - \frac{3}{2} \times (W + F)\right) \times C_{pul},$$ (8.51)

where C_{pul} is the per-unit-length capacitance between the rings which is given as below

$$C_{pul} = \varepsilon_0 \times \varepsilon_{eff} \frac{K\left(\sqrt{1 - k^2}\right)}{K(k)},$$ (8.52)

where ε_{eff} is the effective dielectric constant which is expressed as

$$\varepsilon_{eff} = \frac{\varepsilon_r + 1}{2},$$ (8.53)

where K (k) denotes the complete elliptical integral of the first kind

$$K(k) = \frac{\pi}{2}\sum \left[\frac{(2 \times n)!}{2^{2n} \times (n!)^2}\right]^2 \times \alpha^{2n},$$ (8.54)

With k expressed as

$$k = \frac{F}{F + 2 \times W},$$ (8.55)

The square single ring CSRR is shown in Fig. 8.14 (b), this resonance frequency is determined by the same relation the square double rings CSRR, with the dielectric width between the inner and the outer square is null.

8.3. Results

8.3.1. SIW Filter with Inductive Post-Wall Irises in the C-Band

The substrate used is FR4 substrate the relative permittivity ε_r=4.4, $Tg\delta = 0.015$ and the height h = 0.8 mm. The standard guide of this band having the dimensions a = 40.39 mm and b = 20.19 mm, for the cutoff frequency of TE_{10} Mode $f_{c)10}$=3.71 GHz and according to Eq. 8.1 in TE_{10} mode, the effective width is W_{eff} =19.3 mm.

In order to compare the electromagnetic field distribution in SIW and equivalent rectangular waveguide, the diameter of the metallic via D = 1 mm and the period of the vias P = 1.8 mm. According to Eq. 8.3 the distance between the rows of the centres of via is W_{SIW} =19.9 mm. The length of SIW guide is L_{SIW}=L_{eff}=80 mm, because the term (D^2/ (0.95*P)) does not bring significant change in the propagation phenomenon.

The SIW guide and the equivalent rectangular waveguide are analyzed by HFSS. Fig. 8.16 shows the similarity of the electric field distribution of the TE_{10} mode guided in the two structures.

The comparison the electric field distribution of the TE_{10} mode guided in the SIW guide and in the equivalent rectangular waveguide shows that the dominant mode of the SIW is quite similar to the TE_{10} mode of the conventional waveguide.

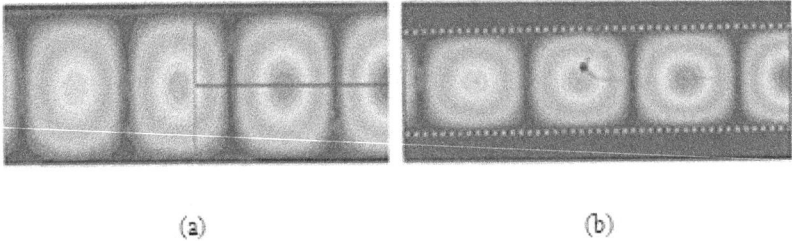

(a)   (b)

Fig. 8.16. Electric field distribution of the TE_{10} mode to the frequency f = 5 GHz for (a) Equivalent rectangular waveguide; (b) SIW Guide.

Now, the SIW structure dimensions are found, a microstrip transition is used to interconnect and match the impedance between a 50 Ω microstrip line and the SIW. The 50 Ω microstrip line, in which the dominant mode is quasi-TEM, can excite well the dominant mode TE_{10} of the SIW, as their electric field distributions are approximate in the profile of the structure. The parameters (the width W_M) of the microstrip line and (the width W_T and the length L_T) the transitions are determined from following formulas (Eq. 8.9, Eq. 8.15, Eq. 8.16, Eq. 8.17 and Eq. 8.18). The microstrip line: W_M=1.4 mm and the dimensions of transition: W_T =7.56 mm and L_T =21.45 mm. A schematic view of a SIW with two stepped transitions is shown in Fig. 8.17.

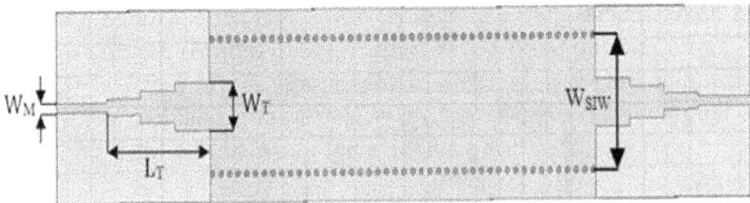

Fig. 8.17. SIW with two stepped transitions.

As shown in Fig. 8.17, the transition comprises three tapers. The dimensions of the tapers as a function of L_T and W_T are summarized in Table 8.1.

172

Fig. 8.18 illustrated the reflection coefficient S11 and the transmission coefficient S21 of SIW with two stepped transitions in the band [4-7] GHz.

Table 8.1. Geometric parameters of the microstrip to SIW planar stepped transition.

Taper i	1	2	3
Width W_{Ti} (mm)	$W_T/3$	$2W_T/3$	W_T
Length L_{Ti} (mm)	$L_T/3$	$L_T/3$	$L_T/3$

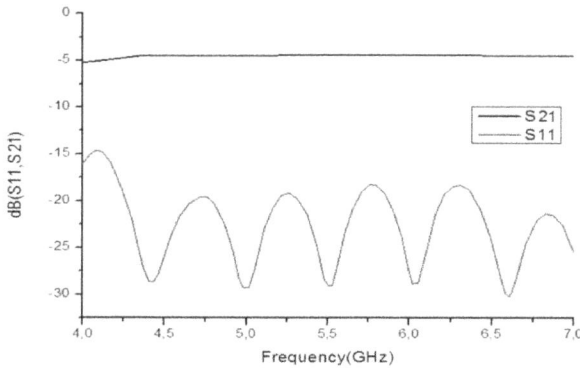

Fig. 8.18. Frequency response of SIW with two stepped transitions.

The results illustrated in Fig. 8.18, indicate that the reflection coefficient S11 remains below -15 dB on the entire band [4-7] and the transmission coefficient S21 is around 5 dB across the entire band.

Before passing to the design of SIW filter iris in the C-band, studying their equivalent circuit using software 2D simulation (ADS). This filter has a central frequency f_0= 5.245 GHz, the absolute bandwidth 730 MHz and the relative bandwidth FBW = 14 %, the ripple is 0.01 dB. Outside the bandwidth, the filter must submit a rejection of -30 dB at 6.5 GHz. By exploiting (Eq. 8.21 and Eq. 8.22) for deduct a 4rd order filter. The circuit model of microwave filter in ADS is shown in Fig. 8.19.

In the Fig. 8.19 $K_{0.1}$ and $K_{4.5}$ correspond to access coupling coefficients, $K_{1.2}$, $K_{2.3}$ and $K_{3.4}$ correspond the coupling coefficients between resonators.

Note that the capacitance and inductance of the same resonator are mutually related. Because what is a model representing the cavity SIW, the choice of their value can be arbitrarily made. By fixing $L = C$ and from following formulas (Eq. 8.23, Eq. 8.25, Eq. 8.26, Eq. 8.27, Eq. 8.28, Eq. 8.29, Eq. 8.30, Eq. 8.31, Eq. 8.32, Eq. 8.33, Eq. 8.34, Eq. 8.35 and Eq. 8.36), you obtain the values of the impedance inverters $K_{i,i+1(0\leq i \leq 4)}$. The function tuning of ADS is used to enable adjustment of the coupling coefficients $K_{i,i+1(0\leq i \leq 4)}$. Finally, the coupling matrix is shown in Fig. 8.20.

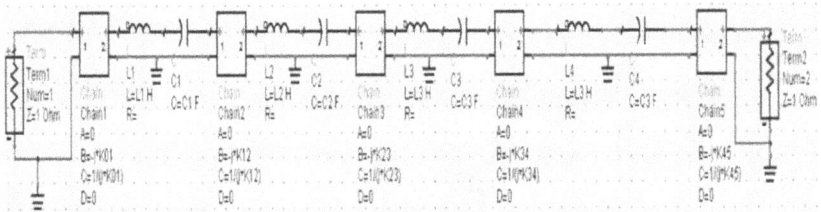

Fig. 8.19. Circuit model of microwave bandpass filter of order 4 in ADS.

	S	1	2	3	4	L
S	0	0.398	0	0	0	0
1	0.398	0	0.106	0	0	0
2	0	0.106	0	0.074	0	0
3	0	0	0.074	0	0.106	0
4	0	0	0	0.106	0	0.398
L	0	0	0	0	0.398	0

Fig. 8.20. Coupling matrix of the microwave bandpass filter of order 4.

The indices S and L correspond to the source and to the load respectively, in other words to the accesses. The indices going from 1 to 4 correspond to the four resonators. The ideal frequency response circuit in the K-band is shown in Fig. 8.21.

The results illustrated in Fig. 8.21, show the filter has a center frequency $f_0 = 5.245$ GHz, the absolute bandwidth 730 MHz and the relative bandwidth FBW= 14 %. The return loss is better than 20 dB between 5.02 GHz and 5.48 GHz. The results of filter respects well the tender specifications.

After the study of equivalent circuit of SIW filter iris in the C-band, passing to the design, by using the FR4 substrate for comparing with the results in [5]. The structure of SIW bandpass filter in the C-band with iris topology a 4rd order is shown in Fig. 8.22.

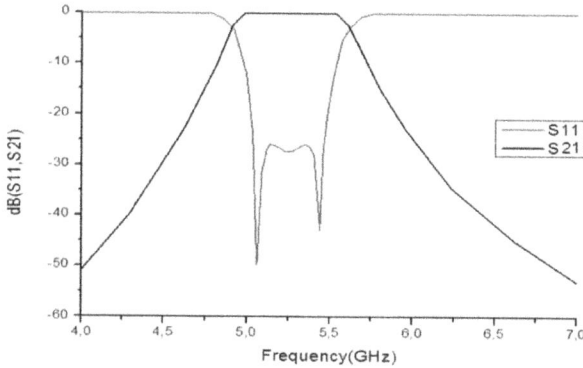

Fig. 8.21. Frequency response of microwave bandpass filter.

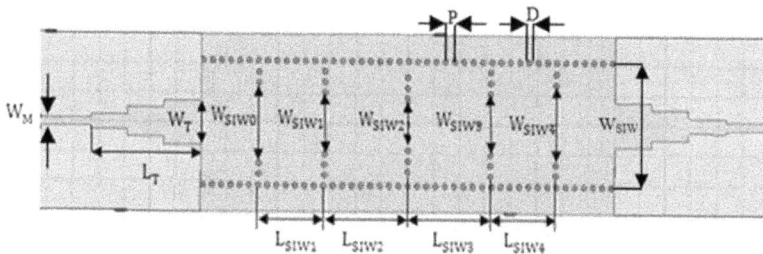

Fig. 8.22. SIW filter with stepped transition using post-wall irises.

In the Fig. 8.22 W_{SIWi} (i=0,1,2,3,4) are the widths of the resonators and L_{SIWi} (i=1,2,3,4) are the lengths of the resonators. Using method the abacus of the estimated width in rectangular waveguide [14] and the following formulas (Eq. 8.37, Eq. 8.38, Eq. 8.39, Eq. 8.40, Eq. 8.41, Eq. 8.42, Eq. 8.43, Eq. 8.44 and Eq. 8.45) to find the widths and the lengths of the resonators, The final dimensions of the structure are optimized by HFSS: $W_{SIW0}= W_{SIW4}$=13.5 mm, $W_{SIW1}= W_{SIW3}$=10.5 mm, W_{SIW2}=9.18 mm, $L_{SIW1}= L_{SIW4}$ =12.7 mm and $L_{SIW2}= L_{SIW3}$=16 mm.

With: W_T=7.56 mm, L_T=21.45 mm, D = 1 mm, P = 1.8 mm, W_{SIW} = 19.9 mm and W_M=1.4 mm.

Fig. 8.23 illustrated the reflection coefficient S11 and the transmission coefficient S21 of SIW bandpass filter iris in the C-band with two stepped transitions and also the results in [5].

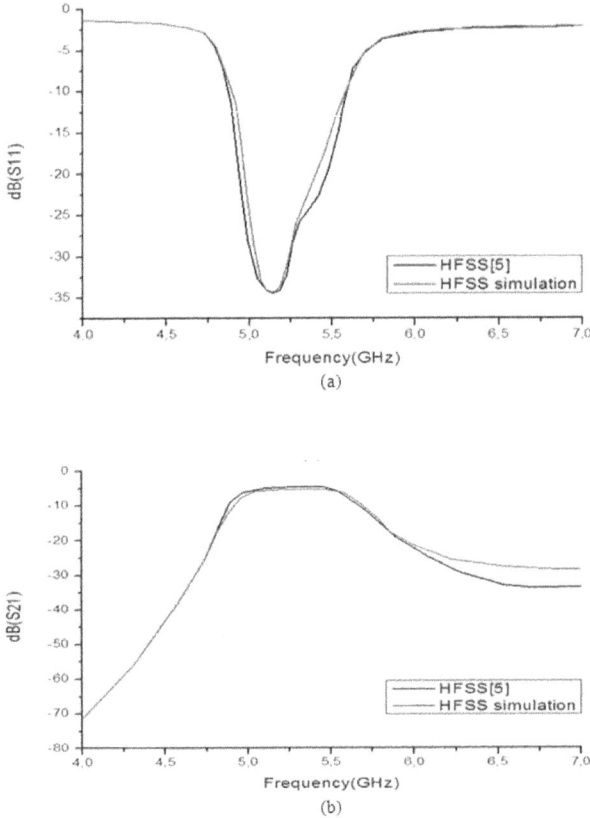

Fig. 8.23. Frequency response of the SIW bandpass filter iris in the C-band with two stepped transitions: (a) Reflection coefficient S11 as a function of frequency; (b) Transmission coefficient S21 as a function of frequency.

The result of the simulation HFSS shows that the center frequency f_0= 5.24 GHz, the absolute bandwidth 730 MHz and the relative bandwidth FBW= 13.9 %. The insertion losses around 5.24 GHz is approximately 5 dB and the return loss is better than 20 dB between 4.95 GHz and 5.35 GHz. The results obtained by HFSS simulation are in good agreement with the results of the reference [5].

8.3.2. SIW Filter with Circular Inductive Post in the C-Band

By using the same FR4 substrate relative permittivity $\varepsilon r = 4.4$, $Tg\delta = 0.015$ and the height $h = 0.8$ mm, with the same diameter of the metallic via $D = 1$ mm and the period of the vias $P = 1.8$ mm and even template filtering function proposed for the SIW filter with inductive post-wall irises in the C-band, for compare with the results in [5].

The SIW filter a 4rd order; the cavities of the filter using inductive post are shown in Fig. 8.24.

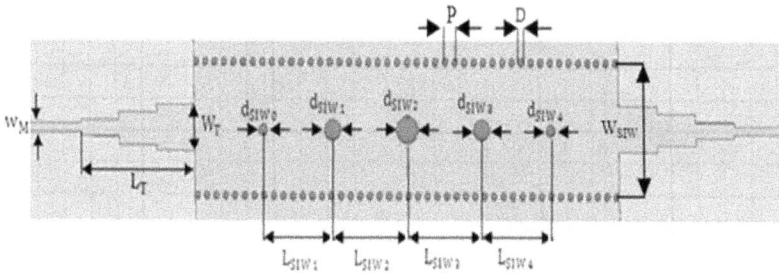

Fig. 8.24. SIW filter with two stepped transitions using inductive post.

In the Fig. 8.24 d_{SIWi} (i=0,1,2,3,4) are the diameter of circular posts and L_{SIWi} (i=1,2,3,4) are the lengths of the resonators. Using method the abacus of the estimated diameter in rectangular waveguide [13, 14] and the following formulas (Eq. 8.37, Eq. 8.38, Eq. 8.39, Eq. 8.40, Eq. 8.41, Eq. 8.42, Eq. 8.43, Eq. 8.44 and Eq. 8.45), you finding the final dimensions of the structure are: $d_{SIW0} = d_{SIW4} = 0.07$ mm, $d_{SIW1} = d_{SIW3} = 0.5$ mm, $d_{SIW2}=0.76$ mm. $L_{SIW1}= L_{SIW4} =13$ mm and $L_{SIW2}= L_{SIW3} = 14$ mm.

With: $W_T=7.56$ mm, $L_T=21.45$ mm, $D = 1$ mm, $P = 1.8$ mm, $W_{SIW} = 19.9$ mm, $W_M=1.4$ mm

Fig. 8.25 illustrated the reflection coefficient S11 and the transmission coefficient S21 of the SIW bandpass filter with two stepped transitions using inductive post in the C-band and also the results in [5].

The result of the simulation HFSS shows that the center frequency $f_0= 5.4$ GHz, the absolute bandwidth 1.2 GHz and the relative bandwidth FBW= 22 %. The insertion loss around 5.4 GHz is approximately 4.7 dB, the return loss is better than 20 dB between 4.8 GHz and 5.92 GHz.

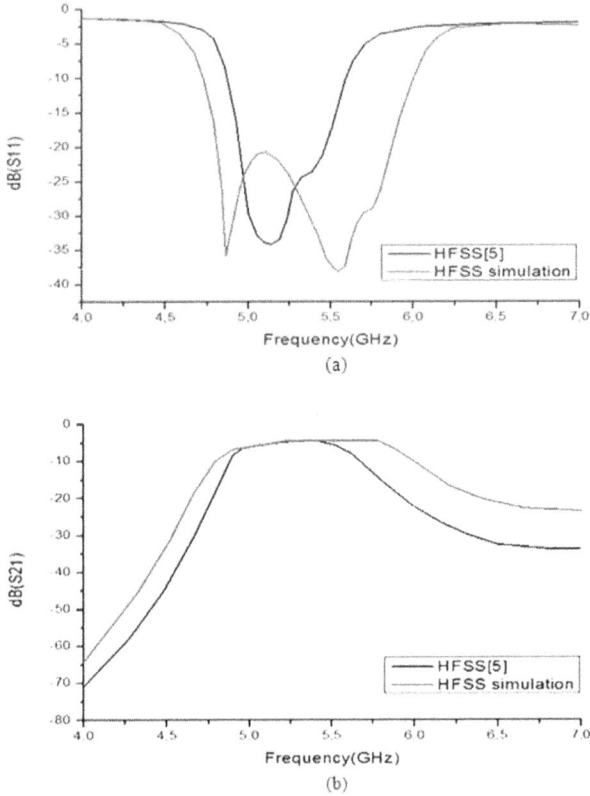

Fig. 8.25. Frequency responses of SIW bandpass filter inductive post with two stepped transitions in the C-band: (a) Reflection coefficient S11 as a function of frequency; (b) Transmission coefficient S21 as a function of frequency.

8.3.3. SIW Filter with Circular Inductive Post in the Q-Band

This filter is designed with a RT/Duroid 5880 substrate the relative permittivity $\varepsilon r = 2.2$ and the height h = 0.254 mm, the diameter of the metallic via D = 0.25 mm and the period of the vias P = 0.4 mm. By following the same approach of design the SIW with transitions, you finding: W_{eff} =3.84 mm, W_{SIW} =4 mm, W_M=0.63, W_T =1.5 mm and L_T =4 mm. A schematic view of a SIW with two tapered transitions is shown in Fig. 8.26.

Fig. 8.27 illustrated the reflection coefficient S11 and the transmission coefficient S21 of SIW with two tapered transitions in the band [33-50] GHz.

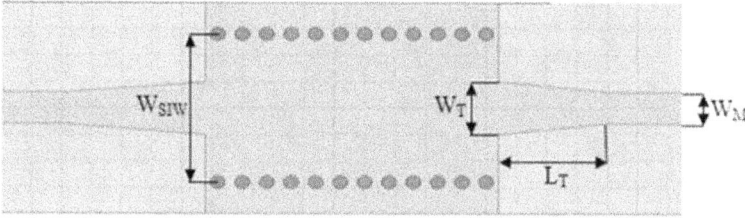

Fig. 8.26. SIW with two tapered transitions.

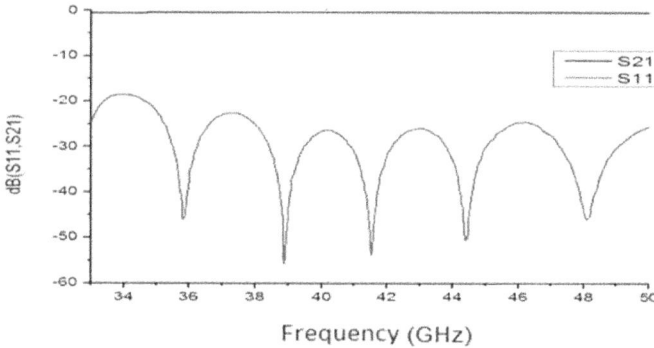

Fig. 8.27. Frequency response of SIW with two tapered transitions.

The results illustrated in Fig. 8.27, indicate that the reflection coefficient S11 remains below -15 dB on the entire band [33-50] GHz and the transmission coefficient S21 is around 0.5 dB across the entire band.

The filtering function of the template proposed for this filter is composed of a central frequency $f_0 = 41.7$ GHz, the absolute bandwidth 1.5 GHz and the relative bandwidth FBW = 3.5 %. The ripple is 0.01 dB. Outside the bandwidth, the filter must submit a rejection of -20 dB at 43.5 GHz, you deduct a 3rd order filter. The circuit model of microwave filter in ADS is shown in Fig. 8.28.

The ADS tuning tool will be called upon to adjust the various settings that are the coupling coefficients and the parallel reactors. Finally, the coupling matrix is shown in Fig. 8.29.

Fig. 8.30 illustrated the reflection coefficient S11 and the transmission coefficient S21 of this filter in the band [33-50] GHz.

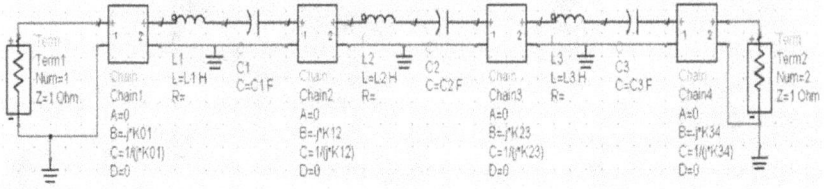

Fig. 8.28. Circuit model of microwave bandpass filter of order 3 in ADS.

	S	1	2	3	L
S	0	0.225	0	0	0
1	0.225	0	0.028	0	0
2	0	0.028	0	0.028	0
3	0	0	0.028	0	0.225
L	0	0	0	0.225	0

Fig. 8.29. Coupling matrix of the microwave bandpass filter of order 3.

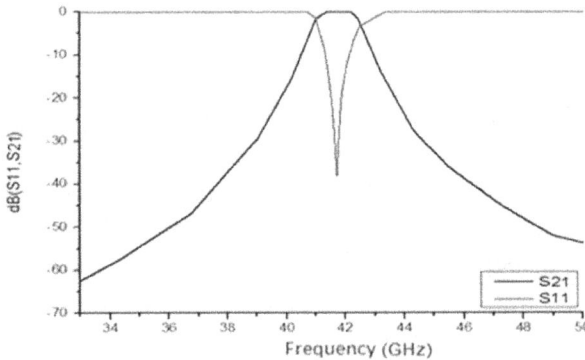

Fig. 8.30. Frequency response of microwave bandpass filter.

The results illustrated in Fig. 8.30, show the filter has a central frequency f_0= 41.7 GHz, the absolute bandwidth 1.5 GHz and the relative bandwidth FBW= 3.5 %. The return loss is better than 20 between 41.55 GHz and 41.85 GHz. The results of filter respects well the tender specifications.

The conception of SIW filter circular inductive post in the band [33-50] GHz uses RT/Duroid 5880 substrate for comparing with the results in [12]. The structure of this filter a 3rd order is shown in Fig. 8.31.

180

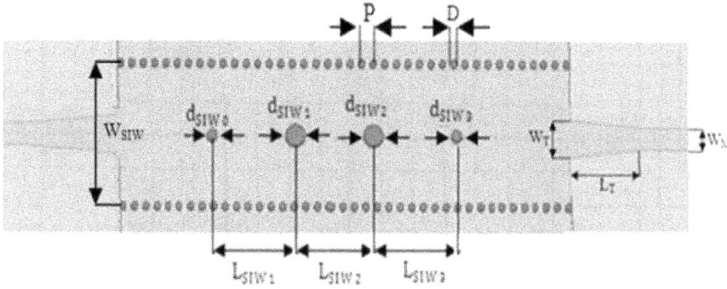

Fig. 8.31. SIW filter with two tapered transitions using inductive post.

In the Fig. 8.31 d_{SIWi} (i=0,1,2,3) are the diameter of circular posts and L_{SIWi} (i=1,2,3) are the lengths of the resonators. By following the same approach, the final dimensions of the structure are: $L_{SIW1}=L_{SIW3}=3$ mm, $L_{SIW2}=3.4$ mm, $d_{SIW0}=d_{SIW3}=0.3$ mm, $d_{SIW1}=d_{SIW2}=0.82$ mm.

With: $W_T=1.5$ mm, $L_T=4$ mm, $D = 0.25$ mm, $P = 0.4$ mm, $W_{SIW} = 4$ mm and $W_M =0.6$ mm.

Fig. 8.32 illustrated the reflection coefficient S11 and the transmission coefficient S21 of the SIW bandpass filter inductive post in the Q-band with two tapered transitions and also the results in [12].

The result of the simulation HFSS shows that the central frequency $f_0 = 41.7$ GHz, the absolute bandwidth 1.5 GHz and the relative bandwidth FBW = 3.5 %. The insertion loss around 41.8 GHz is approximately 0.4 dB, the return loss is better than 15 dB between 41.1 GHz and 42.4 GHz. The results obtained by HFSS simulation are in good agreement with the results in [12].

8.3.4. SIW Filter with Inductive Post-Wall Irises in the V-Band

This filter is designed with a Neltec NY9217 (IM) 5880 substrate the relative permittivity $\varepsilon_r =2.17$ and the height h = 0.2 mm, the diameter of the metallic via D = 0.25 mm and the period of the vias P = 0.4 mm. By following the same approach of design the SIW with two tapered transitions (Fig. 8.26), you finding: $W_{eff} =2.58$ mm, $W_{SIW} =2.8$ mm. $W_M=0.63$ mm, $W_T =1$ mm and $L_T =1.8$ mm.

Fig. 8.33 illustrated the reflection coefficient S11 and the transmission coefficient S21 of SIW with two tapered transitions in the band [50-75] GHz.

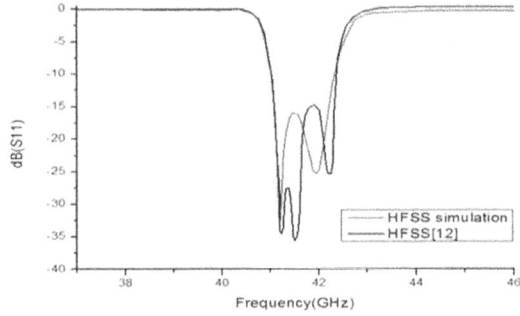

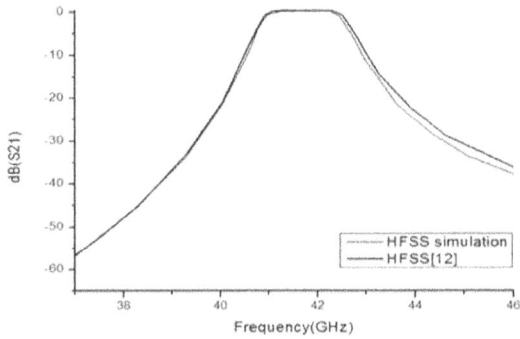

Fig. 8.32. Frequency responses of SIW bandpass filter inductive post in the Q-band with two tapered transitions: (a) Reflection coefficient S11 as a function of frequency; (b) Transmission coefficient S21 as a function of frequency.

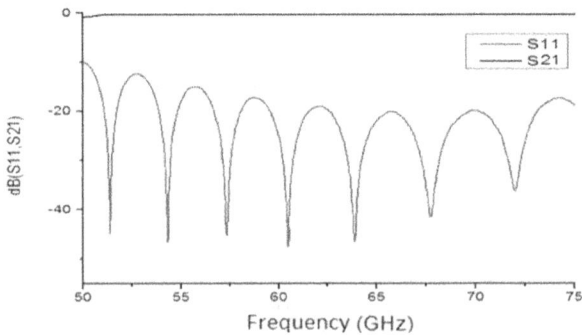

Fig. 8.33. Frequency response of SIW with two tapered transitions.

The results illustrated in Fig. 8.33, indicate that the reflection coefficient S11 remains below -15 dB over 90 % of the band [50-75] GHz and the transmission coefficient S21 is around 0.7 dB across the entire band.

However, the design of SIW filter iris in the V-band has a centre frequency $f_0 = 62$ GHz, the absolute bandwidth 12 GHz and the relative bandwidth FBW = 19 %. The ripple is 0.01 dB. Outside the bandwidth, the filter must submit a rejection of -15 dB at 66.3 GHz. you deduct a 3rd order filter as in Fig. 8.28, the coupling matrix is shown in Fig. 8.34.

	S	1	2	3	L
S	0	0.389	0	0	0
1	0.389	0	0.1255	0	0
2	0	0.1255	0	0.1255	0
3	0	0	0.1255	0	0.389
L	0	0	0	0.389	0

Fig. 8.34. Coupling matrix of the microwave bandpass filter of order 3.

After simulation, the ideal frequency response of the circuit on the band [50-75] GHz as shown in Fig. 8.35.

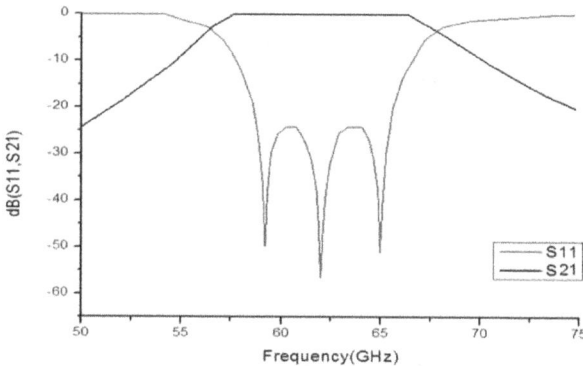

Fig. 8.35. Frequency response of microwave bandpass filter.

The results illustrated in Fig. 8.35, show the filter has a centre frequency $f_0 = 62$ GHz, the absolute bandwidth 12 GHz and the relative bandwidth

FBW= 19 %. The return loss is better than 20 between 58.59 GHz and 65.61 GHz. The results of filter respects well the tender specifications.

After the study of equivalent circuit of SIW filter iris in the band [50-75] GHz, The conception of this filter uses RT/Duroid 5880 Neltec NY9217 (IM) substrate for comparing with the results in [9]. The structure is shown in Fig. 8.36.

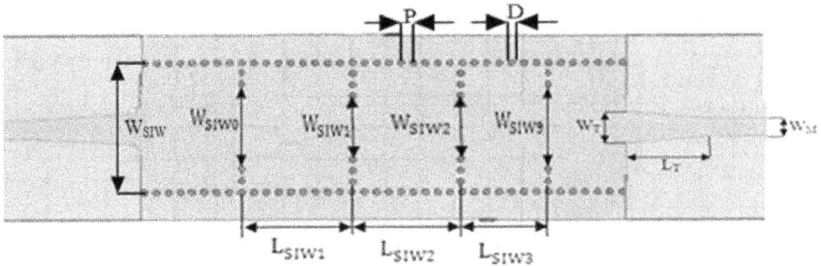

Fig. 8.36. SIW filter with tapered transition using post-wall irises.

The final dimensions of the structure are: $L_{SIW1}=L_{SIW3}=1.65$ mm, $L_{SIW2}=1.8$ mm, $W_{SIW0}=W_{SIW3}=1.91$ mm, $W_{SIW1}=W_{SIW2}=1.51$ mm, $W_T=1$ mm and $L_T=1.8$ mm, with D = 0.25 mm, P = 0.4 mm, W_{SIW} = 2.8 mm, W_M =0.63 mm.

Fig. 8.37 illustrated the reflection coefficient S11 and the transmission coefficient S21 of SIW bandpass filter iris in the V-band with two tapered transitions and also the results in [9].

The result of the simulation HFSS shows that the centre frequency f_0= 62 GHz, the absolute bandwidth 12 GHz and the relative bandwidth FBW= 19 %. The insertion loss around 62 GHz is approximately 0.35 dB, the return loss is better than 15 dB between 60 GHz and 66.3 GHz. The results obtained by HFSS simulation are in good agreement with the results in [9].

8.3.5. SIW Bandpass Filter Based on Metamaterials (CSRRs) in the X-Band

This filter is designed with a RT/Duroid 5880 substrate the relative permittivity εr =2.2 and the height h = 0.254 mm, the diameter of the metallic via D = 0.8 mm and the period of the vias P = 1.6 mm.

Dimensions the SIW with two tapered transitions (Fig. 8.26) are: W_{eff} =15.41 mm, W_{SIW} =16 mm, W_M=0.8, W_T=5.2 mm and L_T=14 mm.

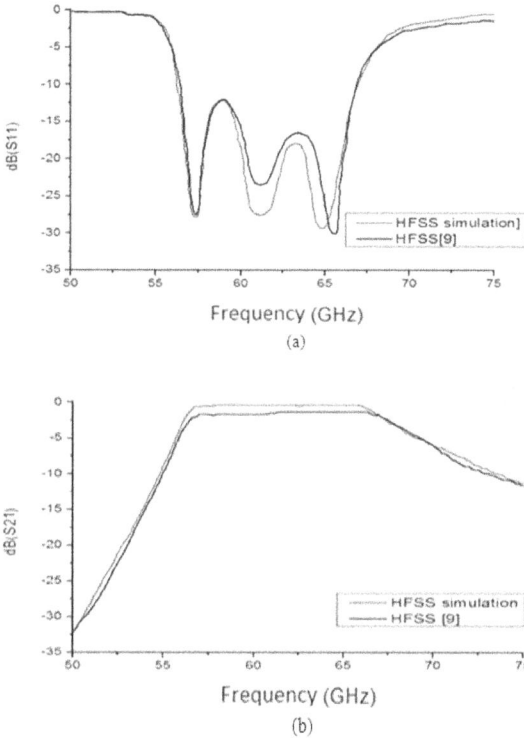

Fig. 8.37. Frequency response of the SIW bandpass filter iris in the V-band with two tapered transitions: (a) Reflection coefficient S11 as a function of frequency; (b) Transmission coefficient S21 as a function of frequency.

Fig. 8.38 illustrated the reflection coefficient S11 and the transmission coefficient S21 of SIW with two tapered transitions in the band [8.2-12.4] GHz.

The results illustrated in Fig. 8.38, indicate that the reflection coefficient S11 remains below-15 dB over 32 % of the frequency band and the transmission coefficient S21 is around 0.9 dB across the entire band.

The SIW bandpass filter based on metamaterials (CSRRs) is designed by two methods.

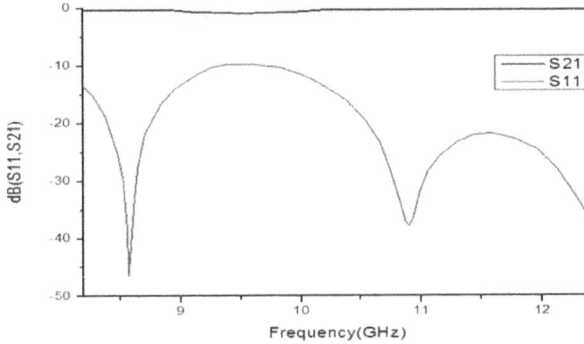

Fig. 8.38. Transmission coefficient S21 and reflection coefficient S11
of SIW with two tapered transitions.

The first method concerns the use of three square single ring CSRRs cells
are etched in the top plane of the SIW.

Before beginning the study of the proposed filter, the resonant properties
of square single ring CSRR are carefully studied starting with a single
CSRR etching in the top plane of the SIW is shown in Fig. 8.39.

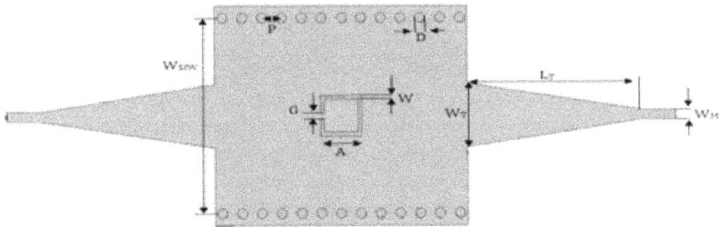

Fig. 8.39. SIW guide with square single ring CSRR.

The dimensions of the CSRR structure are chosen to have resonant
frequency at 10.78 GHz. They are A = 3 mm, W = 0.3 mm and
G = 0.4 mm for a RT/Duroid 5880 substrate (dielectric constant εr= 2.22,
thickness h = 0.254 mm and tan δ = 0.002). The CSRR structure is etched
in the top plane of the SIW exactly the center of a conductor plane of the
SIW guide as shown in Fig. 8.39.

The characteristics of square single ring CSRR on the surface of the
substrate integrated waveguide are studied by varying the value of one
parameter while the rest of the dimensions are being kept constant. The

186

analyses of the performances are based on the central stopband frequency, attenuation, passband frequency, fractional bandwidth and insertion loss.

First case:

The dimensions of the SIW guide and the transition are: $D = 0.8$ mm, $P=1.6$ mm, $W_{SIW} =16$ mm, $W_M=0.8$, $W_T=5.2$ mm and $L_T=14$ mm. The dimensions of square single ring CSRR are: $W = 0.3$ mm and $G = 0.4$ mm. With the length of the side of the square "A" is varied in three different values (A = 3 mm, A = 3.1 mm and A=3.2 mm).

Fig. 8.40 shows the transmission coefficient S21 and the reflection coefficient S11 of the SIW guide with square single ring CSRR in three different values of the length of the side of the square "A".

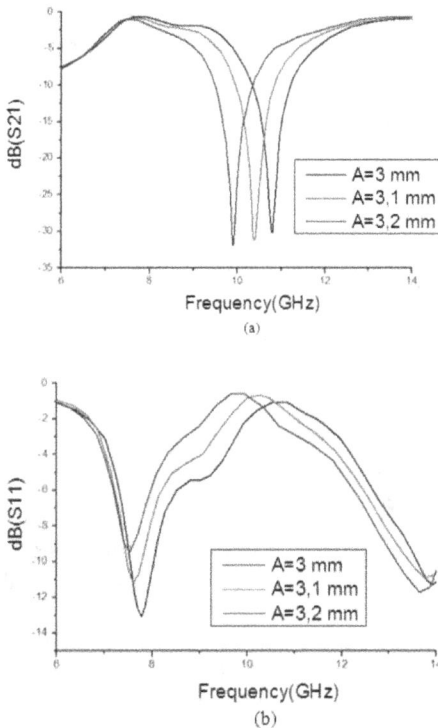

Fig. 8.40. Frequency response of the SIW guide with square single ring CSRR:
(a) Transmission coefficient S21 as a function of frequency;
(b) Reflection coefficient S11 as a function of frequency.

The results simulated by HFSS of the SIW guide with square single ring CSRR in three different values of the length of the side of the square "A" are shown in Table 8.2.

Table 8.2. Frequency response of variation of "A" for SIW guide with square single ring CSRR with dimensions D = 0.8 mm, P = 1.6 mm, W_{SIW} =16 mm, W_M=0.8 mm, W_T=5.2 mm, L_T=14 mm, W = 0.3 mm and G = 0.4 mm.

A (mm)	Centre stopband frequency (GHz)	Attenuation (dB)	Passband frequency (GHz)	Fractional bandwidth	Insertion loss (dB)
3	10.78	-30.3	7 to 9.71	32.4%	1.54
3.1	10.38	-31.4	6.88 to 9.35	30.4%	1.57
3.2	9.9	-32.27	6.8 to 8.8	25.6%	1.58

The results show that an increase in the length of the side of the square "A" a resulted in a decrease in the centre stopband frequency with the increased attenuation, passband band-width is decreased and insertion loss is increased.

Second case:

The dimensions of the SIW guide with square single ring CSRR are: D = 0.8 mm, P = 1.6 mm, W_M=0.8, W_T=5.2 mm, L_T=14 mm, A=3.2 mm, W=0.3 mm and G=0.4 mm. With the width SIW guide "W_{SIW}" is varied in three different values (W_{SIW}=16 mm, W_{SIW}=15 mm and W_{SIW}=14 mm).

Fig. 8.41 shows the transmission coefficient S21 and the reflection coefficient S11 of SIW guide with square single ring CSRR in three different values of the width SIW guide "W_{SIW}".

The results of SIW guide with square single ring CSRR in three different values of the width SIW guide "W_{SIW}" are shown in Table 8.3.

The results show that a decrease in the width SIW guide "W_{SIW}" a resulted in a decrease in passband bandwidth with increased Insertion loss, the centre stopband frequency is fixed and attenuation is increased.

Third case:

The dimensions of the SIW guide with square single ring CSRR are: D = 0.8 mm, P = 1.6 mm, W_{SIW} =14 mm, W_M=0.8, A=3.2 mm, W=0.3 mm and G=0.4 mm. With the dimensions of transition are varied in two values. (L_T=14 mm, W_T =5.2 mm) and (L_T=5.5 mm, W_T =2 mm)

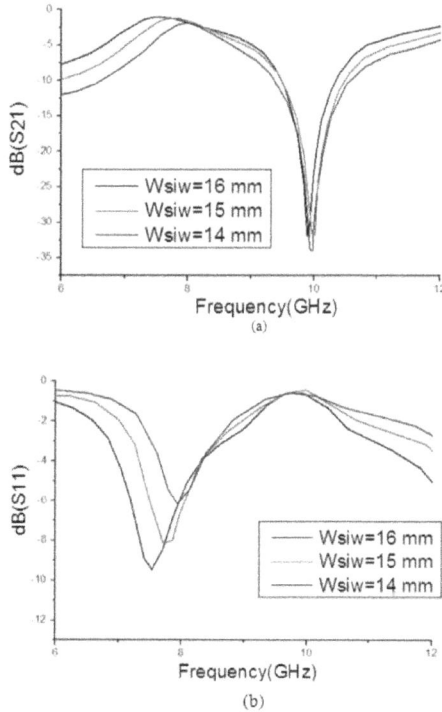

Fig. 8.41. Frequency response of the SIW guide with square single ring CSRR: (a) Transmission coefficient S21 as a function of frequency; (b) Reflection coefficient S11 as a function of frequency.

Table 8.3. Frequency response of variation of "W_{SIW}" for SIW guide with square single ring CSRR with dimensions D = 0.8 mm, P = 1.6 mm, W_M=0.8 mm, W_T=5.2 mm, L_T =14 mm, A=3.2mm, W = 0.3 mm and G = 0.4 mm.

W_{SIW} (mm)	Centre stopband frequency (GHz)	Attenuation (dB)	Passband frequency (GHz)	Fractional bandwidth	Insertion loss (dB)
16	9.9	-32.27	6.8 to 8.88	25.6 %	1.58
15	9.9	-33.23	7.1 to 8.77	21.04 %	1.62
14	9.9	-34.44	7.42 to 8.65	15.3 %	2.1

189

Fig. 8.42 shows the frequency response of SIW guide with square single ring CSRR in two different dimensions of transition (the width L_T and the length W_T).

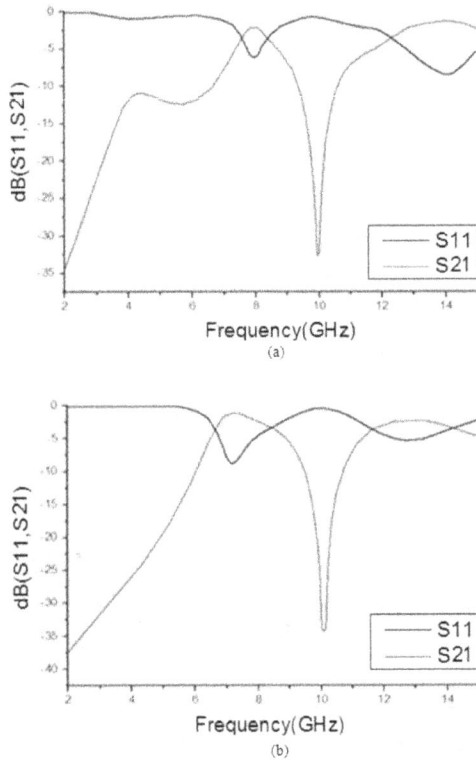

(a)

(b)

Fig. 8.42. (a) Frequency response of the structure (SIW guide with square single ring CSRR) with L_T=14 mm and W_T =5.2 mm; (b) Frequency response of the structure (SIW guide with square single ring CSRR) with L_T=5.5 mm and W_T =2 mm.

The results of SIW guide with square single ring CSRR in two different dimensions of transition (the width L_T and the length W_T) are shown in Table 8.4.

The results show that a decrease in the dimensions of transition a resulted in an increased in passband bandwidth with insertion loss lower, the centre stopband frequency and attenuation are fixed.

Table 8.4. Frequency response of variation the dimensions of transition for SIW guide with square single ring CSRR with dimensions D = 0.8 mm, P = 1.6 mm, W_{SIW} =14 mm, W_M=0.8, A=3.2mm, W=0.3 mm and G=0.4 mm.

(L_T, W_T) (mm)	Centre stopband frequency (GHz)	Attenuation (dB)	Passband frequency (GHz)	Fractional bandwidth	Insertion loss (dB)
(14,5.2)	9.9	-34.44	7.42 to 8.65	15.3 %	2.1
(5.5,2)	9.9	-34.44	6.69 to 8.47	23.7 %	1.6

Finally, the results showed the most important parameters to influence on the central frequency of the stop band attenuation, frequency bandwidth and insertion loss, to achieve a narrow bandwidth with very low insertion loss.

The proposed design of SIW bandpass filter based of three square single ring CSRRs cells is shown in Fig. 8.43.

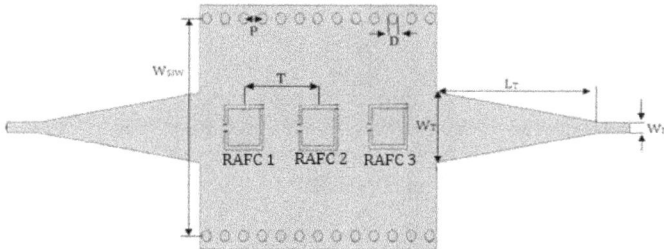

Fig. 8.43. SIW bandpass filter based of three square single ring CSRRs cells.

The dimensions of the SIW guide, the transition and the microstrip line are: D = 0.8 mm, P = 1.6 mm, W_{SIW} =14 mm, W_T =1.72 mm, L_T=5.5 mm, W_M=0.8. The dimensions of square single ring CSRR1 are: A=2.7 mm, W=0.3 mm and G=0.4 mm. The dimensions of square single ring CSRR2 are: A=2.9 mm, W=0.3 mm and G=0.4 mm. The dimensions of square single ring CSRR3 are: A=2.8 mm, W=0.3 mm and G=0.4 mm. The distance between two adjacent square single ring CSRRs is T =7.4 mm.

Fig. 8.44 illustrated the reflection coefficient S11 and the transmission coefficient S21 of SIW bandpass filter based of three square single ring CSRRs cells and also the results in [25].

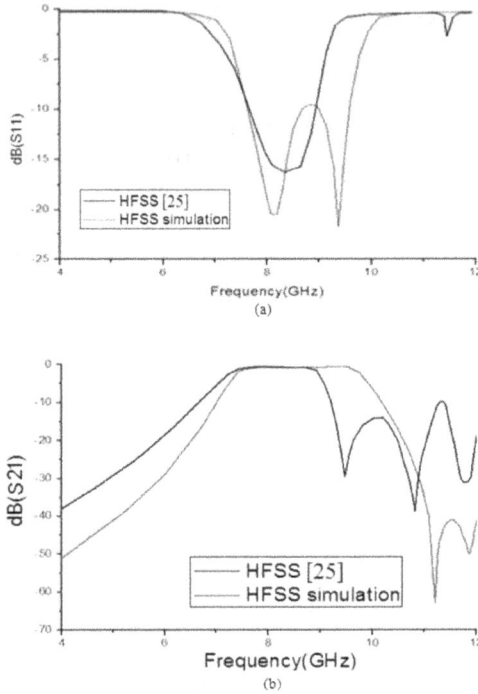

Fig. 8.44. Frequency response of SIW bandpass filter based of three square
single ring CSRRs cells: (a) Reflection coefficient S11 as a function of
frequency; (b) Transmission coefficient S21 as a function of frequency.

The result simulated by HFSS shows that the full frequency passband is
from 7.267 to 9.933 GHz. the center frequency f_0= 8.6 GHz, the absolute
bandwidth 2.67 GHz and the relative bandwidth FBW= 31 %. The
insertion loss around 8.6 GHz is approximately 0.81 dB, the return loss
in the passband is better than 8.91 dB.

On the other side, the result simulated by HFSS in [25] shows that the
full frequency passband is from 7.34 to 9 GHz. the center frequency
f_0= 8.17 GHz, the absolute bandwidth 1.66 GHz and the relative
bandwidth FBW= 20.31 %. The insertion loss around 8.17 GHz is
approximately 0.57 dB, the return loss in the passband is 15 dB.

The results simulated by HFSS in [25] show the frequency responses
better than the results simulated of SIW bandpass filter based of three
square single ring CSRRs cells, owing to the use of resonators CSRRs
more specific about the achievement of narrow bandwidths with very

low insertion loss. To improve the characteristics of the bandpass filter, other resonators CSRSS are used in the second method.

The second method concerns the use of three square double rings CSRRs cells are etched on the top plane of the SIW.

As the first method, the resonant properties of square double ring CSRR are carefully studied starting with a single CSRR etching in the top plane of the SIW is shown in Fig. 8.45.

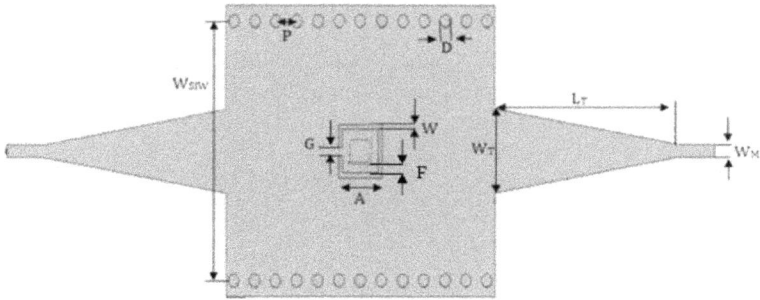

Fig. 8.45. SIW guide with square double rings CSRR.

The dimensions of the CSRR structure are chosen to have resonant frequency at 9.08 GHz. They are A = 3 mm, W = 0.3 mm, G = 0.4 mm and F=0.3 mm for a RT/Duroid 5880 substrate (dielectric constant ε_r= 2.22, thickness h = 0.254 mm and tanδ = 0.002). The CSRR structure is etched in the top plane of the SIW exactly the center of a conductor plane of the SIW guide as shown in Fig. 8.45.

As the first method, the characteristics of square double ring CSRR on the surface of the substrate integrated waveguide are studied by varying the value of one parameter while the rest of the dimensions are being kept constant.

From the results obtained in the first method, the dimensions the SIW, the transition and the microstrip line are: D = 0.8 mm, P = 1.6 mm, W_{SIW}=14 mm, W_T=1.72 mm and L_T=5.5 mm, W_M=0.8.

The dimensions of square double ring CSRR are: W = 0.3 mm, G = 0.4 mm and F=0.3 mm. With the length of the side of the square "A" is varied in three different values (A = 3 mm, A = 2.9 mm and A=2.8 mm).

Fig. 8.46 shows the transmission coefficient S21 and the reflection coefficient S11 of the SIW guide with square double ring CSRR in three different values of the length of the side of the square "A".

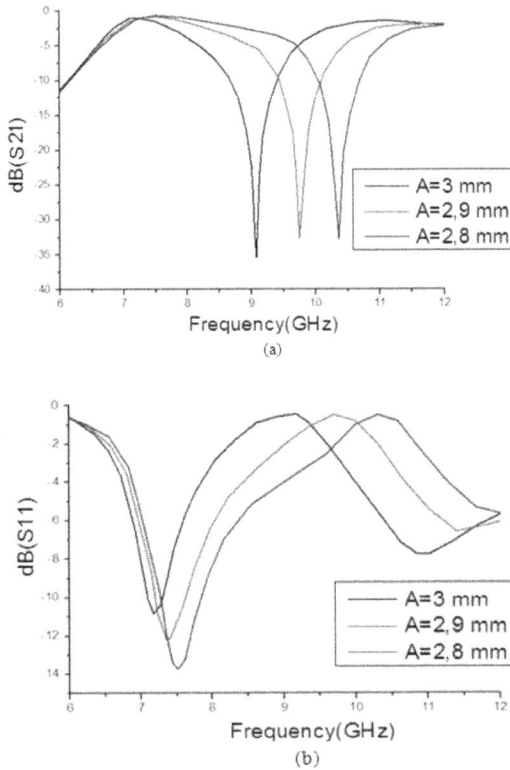

Fig. 8.46. Frequency response of the SIW guide with square double ring CSRR: (a) Transmission coefficient S21 as a function of frequency; (b) Reflection coefficient S11 as a function of frequency.

The results simulated by HFSS of the SIW guide with square double ring CSRR in three different values of the length of the side of the square "A" are shown in Table 8.5.

The results show that a decrease in the length of the side of the square "A" a resulted in an increase in the centre stopband frequency and decreased attenuation, passband bandwidth is increased and insertion loss is decreased.

194

Table 8.5. Frequency response of variation of "A" for SIW guide with square double ring CSRR with dimensions D = 0.8 mm, P = 1.6 mm, W_M=0.8 mm, W_T=1.72 mm, W_{SIW}=14 mm, L_T=5.5 mm, W = 0.3 mm, G = 0.4 mm and F=0.3 mm.

A (mm)	Centre stopband frequency (GHz)	Attenuation (dB)	Passband frequency (GHz)	Fractional bandwidth	Insertion loss (dB)
3	9.08	-37.11	6.66 to 8.06	19.02 %	1.19
2.9	9.73	-35.5	6.7 to 8.7	25.97 %	1.09
2.8	10.33	-33.5	6.8 to 9.4	32 %	0.97

The proposed design of SIW bandpass filter based of three square double ring CSRRs cells is shown in Fig. 8.47.

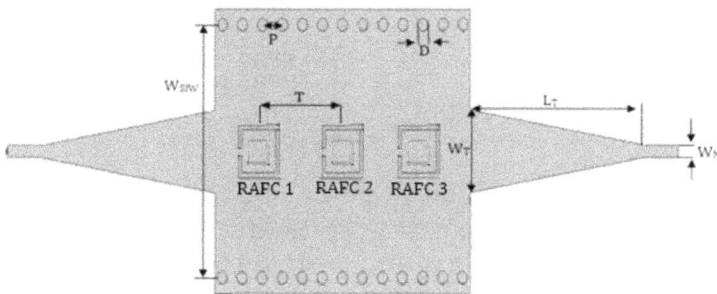

Fig. 8.47. SIW bandpass filter based of three square double ring CSRRs cells.

The dimensions of the SIW guide, the transition and the microstrip line are: D = 0.8 mm, P = 1.6 mm, W_{SIW} =14 mm, W_T =1.72 mm and L_T =5.5 mm, W_M=0.8. The dimensions of square double ring CSRR1 are: A=2.7 mm, W=0.3 mm, G=0.4 mm and F=0.3 mm. The dimensions of square double ring CSRR2 are: A=2.9 mm, W=0.3 mm, G=0.4 mm and F=0.3 mm. The dimensions of square double ring CSRR3 are: A=2.8 mm, W=0.3 mm, G=0.4 mm and F=0.3 mm. The distance between two adjacent square double ring CSRRs is T =9 mm.

Fig. 8.48 illustrated the reflection coefficient S11 and the transmission coefficient S21 of SIW bandpass filter based of three square double ring CSRRs cells and also the results in [25].

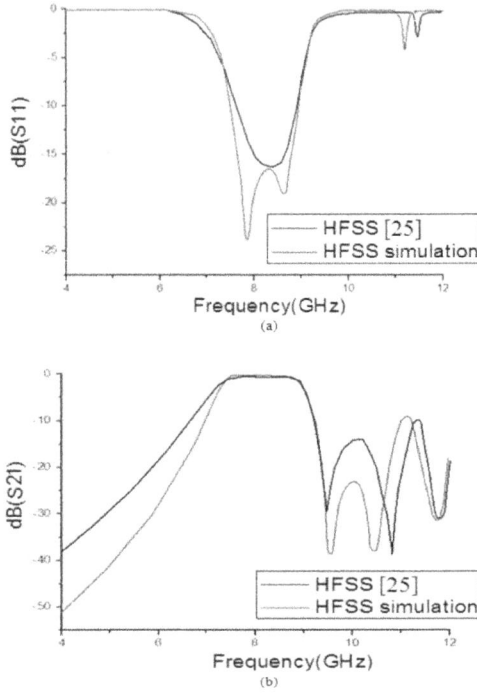

Fig. 8.48. Frequency response of SIW filter based of three square double ring CSRRs cells: (a) Reflection coefficient S11 as a function of frequency; (b) Transmission coefficient S21 as a function of frequency.

The results simulated by HFSS are in good agreement with the results in [25]. The results simulated by HFSS shows that the full frequency passband is from 7.34 to 9 GHz. the center frequency f_0= 8.17 GHz, the absolute bandwidth 1.66 GHz and the relative bandwidth FBW= 20.31%. The insertion loss around 8.17 GHz is approximately 0.57 dB, the return loss in the passband is better than 15 dB.

8.3.6. Design of a SIW Bandpass Filter in the X-Band with Three Methods

The substrate used is Arlon 25N the relative permittivity ε_r=3.38, and the height h = 0.81 mm, the diameter of the metallic via D = 0.8 mm and the period of the vias P = 1.6 mm. Dimensions the SIW with two tapered transitions (Fig. 8.26) are: W_{SIW} =13 mm, W_M=1.8, W_T=5.2 mm and L_T=13 mm.

Fig. 8.49 illustrated the reflection coefficient S11 and the transmission coefficient S21 of SIW with two tapered transitions in the band [8.2-12.4] GHz.

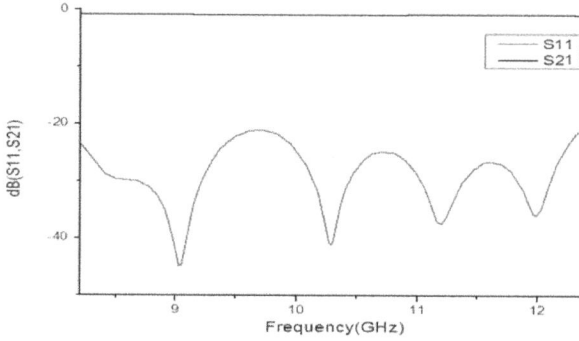

Fig. 8.49. Transmission coefficient S21 and reflection coefficient S11 of SIW with two tapered transitions.

The results illustrated in Fig. 8.49, indicate that the reflection coefficient S11 remains below -20 dB on the entire band [8.2-12.4] and the transmission coefficient S21 is around 1 dB across the entire band.

This filter has a central frequency f_0= 9.675 GHz, the absolute bandwidth 1.13 GHz and the relative bandwidth FBW = 11.68 %, the ripple is 0.01 dB. Outside the bandwidth, the filter must submit a rejection of -15 dB at 11 GHz.

The iris topology and the inductive post topology for SIW filter a 3rd order are shown in Fig. 8.50, with the following dimensions:

L_{SIW1}=L_{SIW3}=9.14 mm, L_{SIW2}=10.2 mm.

W_{SIW0}=W_{SIW3}=8 mm, W_{SIW1}=W_{SIW2}=6 mm.

d_{SIW0}=d_{SIW3}=0.18 mm, d_{SIW1}=d_{SIW2}=1.02 mm.

With: W_T=5.2 mm, L_T=13 mm, D=0.8 mm, P = 1.6 mm, W_{SIW}= 16 mm, W_M=1.8 mm.

The SIW bandpass filter based of three square double ring CSRRs cells is shown in Fig. 8.51, with the following dimensions: D = 0.8 mm, P = 1.6 mm, W_{SIW} = 9.2 mm, W_T = 1.6 mm and L_T = 5.5 mm, W_M= 0.8. The dimensions of square double ring CSRR1 are: A = 2.5 mm, W = 0.3 mm, G = 0.4 mm and F = 0.3 mm. The dimensions of square double ring CSRR2 are: A = 2.6 mm, W = 0.3 mm, G = 0.4 mm and F = 0.3 mm. The dimensions of square double ring CSRR3 are: A=2.5 mm, W=0.3 mm, G=0.4 mm and F=0.3 mm. The distance between two adjacent square double ring CSRRs is T =8.9 mm.

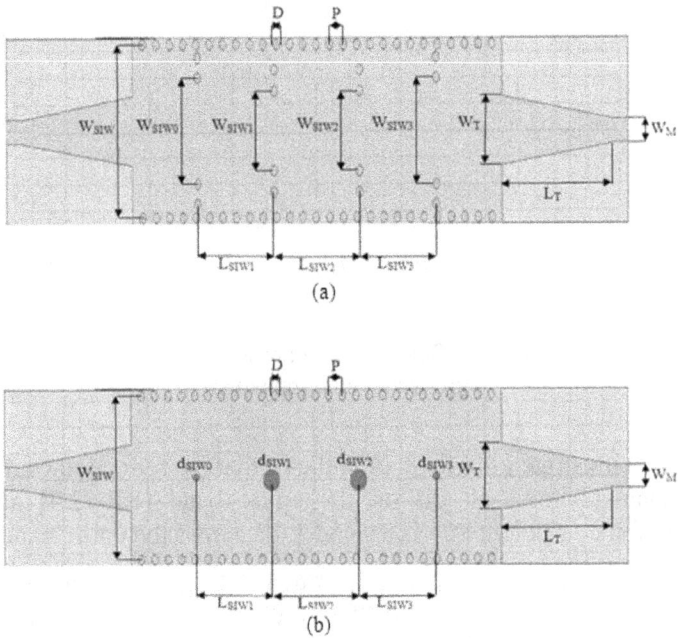

Fig. 8.50. (a) SIW filter with tapered transition using post-wall irises; (b) SIW filter with two tapered transitions using inductive post.

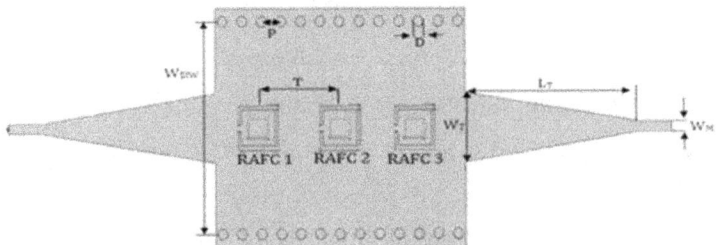

Fig. 8.51. SIW bandpass filter based of three square double ring CSRRs cells.

Fig. 8.52 illustrated the frequency response of SIW bandpass filter in the X-band with iris, inductive post and based of three square double ring CSRRs cells.

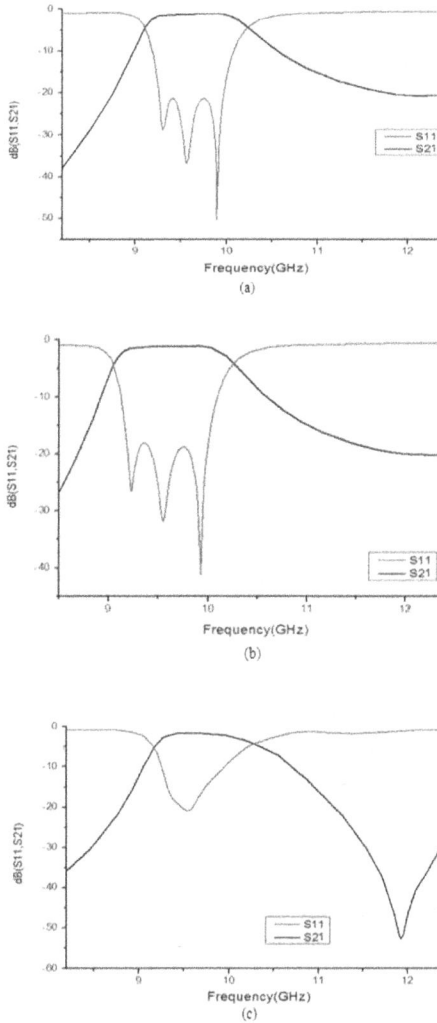

Fig. 8.52. (a) Frequency response of the SIW bandpass filter iris in the X-band with two tapered transitions; (b) Frequency responses of SIW bandpass filter inductive post in the X-band with two tapered transitions; (c) Frequency response of SIW filter based of three square double ring CSRRs cells in the X-band with two tapered transitions.

The results simulated by HFSS are shown in Table 8.6.

Table 8.6. Comparison of SIW filters.

Topologies	Centre frequency (GHz)	Fractional bandwidth	Insertion loss (dB)	Size (mm)
Iris	9.675	11.68 %	1.08	74
Inductive post	9.64	11.70 %	1.08	74
Metamaterials (CSRRs)	9.7	11.37 %	1.26	39

The results show that SIW filter iris and SIW filter inductive post are a return loss in the passband is better than 20 dB, by against SIW bandpass filter based of three square double ring CSRRs cells has a high rejection -52.04 dB to 11.8 GHz frequency, with a small size compared to other filters.

8.4. Conclusions

In this chapter, the way forward for the design of SIW bandpass filters are presented by using three different methods: the iris topology, the inductive post topology and to use the metamaterials (CSRRs). The results show that the SIW filter with iris and with inductive post has good bandwidth adaptation, in contrast the SIW filter based on metamaterials (CSRRs) has an acceptable adaptation with a very small size compared to other filters. The proposed filters are applied in the inter-satellite communications, are easy for integration with other planar circuit compared by using conventional waveguide. The design method is discussed, the results from our analysis are in good agreement with previous research done on this topic. These filters meet the constraints of cost cutting and simplicity of manufacturing imposed by market developments microwave components.

References

[1]. Bouchra Rahali and Mohammed Feham, Design of K-Band substrate integrated waveguide coupler, circulator and power divider, *International Journal of Information and Electronics Engineering,* Vol. 4, Issue 1, 2014, pp. 47–53.

[2]. Ahmed Rhbanou, Seddik Bri, Mohamed Sabbane, Design of C-band substrate integrated waveguide band-pass filter, *European Journal of Scientific Research,* Vol. 123, Issue 3, 2014, pp. 233-245.

[3]. Ahmed Rhbanou, Seddik Br, Design of substrate integrated waveguide pass filter at [33-75] GHz band, *International Journal of Engineering and Technology,* Vol. 6, Issue 6, 2014, pp. 2815-2825.

[4]. Garima Pathak, Substrate integrated waveguide based RF MEMS cavity filter, *International Journal of Recent Technology and Engineering,* Vol. 2, Issue 5, 2013, pp. 46–49.

[5]. Changjun Liu and Kama Huang, A Compact Substrate Integrated Waveguide Band-pass Filter, *PIERS Proceedings,* 2010, pp. 1135–1138.

[6]. B. H. Ahmad, Siti Sabariah Sabri and A. R. Othman, Design of a compact X-Band substrate integrated waveguide directional coupler, *International Journal of Engineering & Technology,* Vol. 5, Issue 2, 2013, pp. 1905–1911.

[7]. Yasser Arfat, Sharad P. Singh, Sandeep Arya, Saleem Khan, Modelling, Design and Parametric Considerations for different Dielectric Materials on Substrate Integrated Waveguide, *WSEAS Transactions on Communications,* Vol. 13, 2014, pp. 94–98.

[8]. Yongmao Huang, Zhenhai Shao and Lianfu Liu, A substrate integrated waveguide bandpass filter using novel defected ground structure shape, *Progress In Electromagnetics Research,* Vol. 135, 2013, pp. 201–213.

[9]. K. Nouri, K. Haddadi, O. Benzaïm, T. Lasri and M. Feham, Substrate integrated waveguide (SIW) inductive window band-pass filter based on post-wall irises, *The European Physical Journal Applied Physics,* Vol. 53, Issue 3, 2011, pp. 33607–33611.

[10]. A. Adabi and M. Tayarani, Substrate integration of dual inductive post waveguide filter, *Progress In Electromagnetics Research B,* Vol. 7, 2008, pp. 321–329.

[11]. A. Ismail, M. S. Razalli, M. A. Mahdi, R. S. A. R. Abdullah N. K. Noordin and M. F. A. Rasid, X-band trisection substrate-integrated waveguide quasi-elliptic filter, *Progress in Electromagnetics Research,* Vol. 85, 2008, pp. 133–145.

[12]. M. N. Husain, G. S. Tan, K. S. Tan, Enhanced Performance of Substrate Integrated Waveguide Bandstop Filter using Circular and Radial Cavity Resonator, *International Journal of Engineering and Technology,* Vol. 6, Issue 2, 2014, pp. 1268–1277.

[13]. Z. Wang, S. Bu, and Z. Luo, A KA-band third-order cross-coupled Substrate Integrated Waveguide Bandstop Filter base on 3D LTCC, *Progress in Electromagnetics Research C,* Vol. 17, 2010, pp. 173–180.

[14]. D. Zelenchuk, V. Fusco, Low insertion loss substrate integrated waveguide quasi-elliptic filters for V-band wireless personal area network applications, *IET Microwaves, Antennas and Propagation,* Vol. 5, Issue 8, 2011, pp. 921–927.

[15]. Arifur Rahman Chowdhury, Mohammad Nurul Hoque, Abdullah Al Mamun, Half-Mode substrate integrated waveguide, *LAP LAMBERT Academic Publishing*, 2011.

[16]. Jia-Sheng Hong and M. J. Lancaster, Microstrip filters for RF/Microwave applications, *John Wiley and Sons, Inc,* 2001.

[17]. Hemendra Kumar, Ruchira Jadhav and Sulabha Ranade, A review on substrate integrated waveguide and its microstrip interconnect, *Journal of Electronics and Communication Engineering,* Vol. 3, Issue 5, 2012, pp. 36–40.

[18]. K. Nouri, M. Feham, Mehdi Damou and Tayeb Habib Chawki Bouazza, Design of substrate integrated waveguide micro-wave planar directional coupler, *International Journal of Scientific & Engineering Research,* Vol. 5, Issue 2, 2014, pp. 1239–1242.

[19]. George Matthaei, Leo Young, E. M. T. Jones, Microwave filters, impedance-matching networks and coupling structures, *Dedham, MA: Artech House,* 1980.

[20]. Nouri Keltouma, Feham Mohammed and Adnan Saghir, Design and characterization of tapered transition and inductive window filter based on Substrate Integrated Waveguide technology (SIW), *International Journal of Computer Science,* Vol. 8, Issue 6, 2011, pp. 135–138.

[21]. Bong S. Kim, Jae W. Lee, Kwang S. Kim, Myung S. Song, PCB substrate integrated waveguide-filter using via fences at millimeter-wave, *IEEE MTT-S Digest,* Vol. 2, 2004, pp. 1097–1100.

[22]. Z. G. Wang, X. Q. Li, S. P. Zhou, B. Yan, R. M. Xu and W. G. Lin, Half mode substrate integrated folded waveguide (HMSIFW) and partial H-plane bandpass filter, *Progress in Electromagnetics Research,* Vol. 101, 2010, pp. 203–216.

[23]. Woon-Gi Yeo, Tae-Yoon Seo, Jae W. Lee, and Choon Sik Cho, H-Plane sectoral filtering horn antenna in PCB substrates using via fences at millimetre-wave, in *Proceedings of the 37th European Microwave Conference,* 2007, pp. 818–821.

[24]. Damou Mehdi, Nouri Keltouma, Taybe Habib Chawki Bouazza and Meghnia Feham, Design of substrat integrated waveguide bandpass filter of SCRRs in the microstrip line, *International Journal of Engineering Research and General Science,* Vol. 2, Issue 3, 2014, pp. 302–314.

[25]. Qun Wu, Ming-Feng Wu, and Fan-Yi Meng, Jian Wu, Le-Wei Li, SRRs' Artificial Magnetic Metamaterials Modeling Using Transmission Line Theory, in *Proceedings of the Progress In Electromagnetics Research Symposium,* Vol. 1, Issue 5, 2005, pp. 630–633.

[26]. F. Martín, J. Bonache, F. Falcone, M. Sorolla and R. Marqués, Split ring resonator-based left-handed coplanar waveguide, *Applied Physics Letters,* Vol. 83, Issue 22, 2003, pp. 4652–4654.

[27]. P. Yasar-Orten, E. Ekmekci and G. Turhan-Sayan, Equivalent circuit models for split-ring resonator arrays, *PIERS Proceedings,* 2010, pp. 534–537.

[28]. G. Siso, M. Gil, M. Aranda, J. Bonache, F. Martin, Miniaturization of Planar Microwave Devices by Means of Complementary Spiral Resonators (CSRs): Design of Quadrature Phase Shifters, *Radioengineering*, Vol. 18, Issue 2, 2009, pp. 144–148.

[29]. X. Lai, Q. Li, P. Y. Qin, B. Wu and C.-H. Liang, A novel wideband bandpass filter based on complementary split-ring resonator, *Progress in Electromagnetics Research C*, Vol. 1, 2008, pp. 177–184.

[30]. H. Bahrami and M. Hakkak, Analysis and design of highly compact bandpass waveguide filter utilizing complementary split ring resonators (CSRR), *Progress in Electromagnetics Research*, Vol. 80, 2008, pp. 107–122.

[31]. M. Mohammed Bait-Suwailam, Miniaturized bandstop filters using slotted-complementary resonators, *International Journal of Digital Information and Wireless Communications (IJDIWC)*, Vol. 4, Issue 3, 2014, pp. 401–407.

[32]. R. S. Kshetrimayum, S. Kallapudi and S. S. Karthikeyan, Stop band characteristics for periodic patterns of CSRRs in the ground plane, *International Journal of Microwave and Optical Technology*, Vol. 2, ssue 3, 2007, pp. 210–215.

[33]. Ahmed Rhbanou, Seddik Bri, Mohamed Sabbane, Design of substrat integrated waveguide bandpass filter based on metamaterials CSRRs, *Electrical and Electronic Engineering*, Vol. 4, Issue 4, 2014, pp. 63-72.

[34]. J. D. Baena, J. Bonache, F. Martin, R. M. Sillero, F. Falcone, T. Lopetegi, M. A. G. Laso, J. Garcia-Garcia, I. Gil, M. F. Portillo, M. Sorolla, Equivalent-circuit models for split-ring resonators and complementary split-ring resonators coupled to planar transmission lines, *IEEE Transactions on Microwave Theory and Techniques*, Vol. 53, Issue 3, 2005, pp. 1451–1461.

[35]. Vidyalakshmi. M. R and Dr. S. Raghavan, Comparison of optimization techniques for square split ring resonator, *International Journal of Microwave and Optical Technology*, Vol. 5, Issue 5, 2010, pp. 280–286.

Index

www.ingramcontent.com/pod-product-compliance
Lightning Source LLC
Chambersburg PA
CBHW050458190326
41458CB00005B/1342